Infrared Transmission Spectra of Carbonate Minerals

THE NATURAL HISTORY MUSEUM

Infrared Transmission Spectra of Carbonate Minerals

G. C. Jones

Department of Mineralogy
The Natural History Museum
London, UK

and

B. Jackson

Department of Geology
Royal Museum of Scotland
Edinburgh, UK

A collaborative project of
The Natural History Museum
and
National Museums of Scotland

SPRINGER-SCIENCE+BUSINESS MEDIA, B.V.

First edition 1993

© 1993 Springer Science+Business Media Dordrecht
Originally published by Chapman & Hall in 1993
Softcover reprint of the hardcover 1st edition 1993

Typeset at the Natural History Museum

ISBN 978-94-010-4940-5   ISBN 978-94-011-2120-0 (eBook)
DOI 10.1007/978-94-011-2120-0

A catalogue record for this book is available from the British Library
Library of Congress Cataloging-in-Publication Data available

∞ Printed on acid-free text paper, manufactured in accordance with
ANSI/NISO Z39.48-1992(Permanence of Paper)

Contents

Foreword

The selected spectra presented in this volume are a testimony to the diversity of mineral carbonates. Their compositional variety embraces many of the chemical elements and is increased by the frequent presence of solid solution between members. They occur in all the broad categories of rock types: igneous, metamorphic, metasomatic and sedimentary; and they are often associated with important ores and rare element deposits. Carbonates are not only of significance in the geological domain, but also in industry and materials science. Accurate identification of the compounds is, therefore, vital for a proper understanding of any carbonate bearing system.

The development of Fourier transform infrared spectrometry has been for some years at the stage where the acquisition of spectra is relatively simple, rapid and with good resolution. For identification, the method is inexpensive and can provide additional information on the nature of the chemical bonding. It is particularly suited to carbonates because of its ability to discriminate clearly between the different members.

It is obvious that to be able to produce a large set of definitive spectra, a source of well-characterized minerals is required, but the location of such a source is not necessarily so obvious. Our two museums – The Natural History Museum in London and the National Museums of Scotland in Edinburgh – have joined forces to provide such a source, using their renowned mineral collections and authenticating each mineral by modern advanced methods of analysis and identification.

This volume is the product of several years' work of high quality. We believe that it gives for today the most readily available compilation of reliable IR spectra as an invaluable reference tool for many.

Professor Paul Henderson
Keeper of Mineralogy
The Natural History Museum, London

and

Dr Ian Rolfe
Keeper of Geology
National Museums of Scotland, Edinburgh

May, 1993

Introduction

The purpose of this compilation is to make available recently-acquired spectra of as many well-characterized carbonate minerals as possible in order to further the use of infrared spectroscopy in mineralogy.

With the recent increased availability of Fourier transform spectrometers for routine laboratory use, there is great potential for infrared spectroscopy to become more widely used, both for the rapid identification of minerals and for more detailed structural studies. Despite being an established analytical technique, mineralogical infrared spectroscopy has been handicapped by a lack of high-quality reference spectra. There is currently no infrared equivalent of the JCPDS Mineral Powder Diffraction File and many new mineral descriptions still lack infrared spectra. Several compilations of mineral spectra are available but are far from comprehensive and are of variable reliability. Published mineral spectra are scattered throughout numerous journals and are often poorly reproduced with limited frequency ranges.

The success of any comparative technique depends to a great extent on the availability and quality of reference standards. A spectrum is a virtually unique "fingerprint" of a material and, accurately reproduced, is a much more useful aid to identification than tabulated absorptions alone, from which all subtleties of detail are lost. The more spectra that are published, the more widely used the technique becomes and so the cycle continues.

The authors acknowledge the co-operation and advice of their colleagues, particularly John Francis and Peter Davidson for x-ray diffraction work, and the mineral curators for their tolerance of our frequent requests for material from their best specimens.

Introduction

A guide to the book

The mineral specimens

The majority of the mineral specimens used in this compilation are from the collections of the Mineralogy Department of The Natural History Museum, London, and the Department of Geology, Royal Museum of Scotland, Edinburgh. Others were acquired from dealers and colleagues specifically for this work.

The criteria used in selecting specimens were as follows:
- purity and homogeneity;
- specimens previously used as x-ray powder diffraction standards where possible;
- ease of contamination-free, unambiguous sampling;
- specimens from type or classic localities.

The rarity of some species has made it necessary to compromise these criteria in a small number of cases. In some cases a relatively common mineral has been omitted from this collection, e.g. natron, thermonatrite, stichtite/barbertonite etc. because, despite sampling many specimens from various localities, no pure or unambiguous material could be separated.

Instrumentation and sample preparation

All spectra were recorded in transmission mode, using a Fourier transform infrared spectrophotometer (Philips PU9800) and potassium bromide pressed disks. The instrument was purged with dry, CO_2-free air and a blank KBr disk was used to generate the background which is automatically subtracted during transformation. 50 scans were acquired using a DTGS detector at a resolution of $2\,cm^{-1}$ followed by two-pass selective smoothing.

Samples were not weighed, but the quantity used was adjusted to give the strongest peak maximum at approximately 20%T or less, without loss of detail around this peak, subject to the availability of material. Most samples were used without prior drying so as to avoid the risk of thermal alteration. Some powdery or poorly-crystalline minerals have significant amounts of adsorbed water which was reduced by allowing the pressed disk to remain in the dry environment of the sample chamber for several hours. This technique was not used for hydrated minerals, some of which can dehydrate very easily to other phases.

The spectra

All spectra were recorded over the frequency range $4400–225\,cm^{-1}$, but as none of the minerals studied had absorption peaks in the range $4400–4000\,cm^{-1}$, the spectra are reproduced here from 4000 to $225\,cm^{-1}$ to make best use of the available format. The spectra also have their vertical expansion adjusted for the same reason. Where multiple sharp peaks are poorly reproduced in the standard format, an expanded wavenumber plot has been included.

The text pages

Name:
    The mineral name in bold type corresponds to that in *Hey's Mineral Index* (Clark (1993)).

Formula, crystal system and space group:
    These data are taken from the reference sources listed at the end of this introduction or from later published work where available.

Mineral group:
    The "mineral group" is that given by *The Mineral Database* (1989) but with alternative groupings shown where these draw attention to relationships between spectra.

Chemical class, chemical type:
    These are taken from Ferraiolo (1982).

Specimen:
    The BM and RMS numbers correspond to registered museum specimens.

    The description corresponds to that on the specimen registration slip (where available), modified as necessary to reflect sampling.

Source:
    This is the locality as recorded on the specimen registration slip, except that some place names have been changed to current usage. Type localities are noted where appropriate.

Spectrum ref. no:
    This is a unique identifier for the spectrum (there may be more than one spectrum per mineral name).

Sample medium:
    This will usually be KBr disk but other techniques may be used where demanded by the nature of the sample.

XRD:
    A number, if given, indicates that the specimen has been examined by x-ray powder diffraction. The suffix (std) indicates the specimen is one that has been used to produce a standard reference diffraction film in The Mineralogy Dept. NHM. Such standards will have been compared with published x-ray data and naturally-occurring, well-characterised mineral specimens. Comparison with the corresponding JCPDS data will also have been made but is not necessarily used as the final criteria for mineral identity.

Composition:
    The chemical composition of most specimens has been checked where possible, using an analytical scanning electron microscope with energy-dispersive x-ray spectrometry facility. Elements with atomic number below that of fluorine are not detectable by this technique, e.g. boron, carbon and oxygen. Fluorine is only detectable when present in major amounts. Ratios quoted are semi-quantitative atomic ratios. Other elements are also listed where present at detectable levels.

Peak Table:
    All spectral data have been obtained via a "peak-pick" program, followed by manual examination and editing to exclude spurious data and include significant shoulders and other diagnostic features. The tables include some peaks that may not be clearly visible on the spectra as reproduced, due to restrictions of the format, they are however visible on expanded plots. The frequencies of these and other minor features are shown in normal type, the major features in bold, as an aid in relating the peak table to the spectrum. Frequencies in square brackets are due to adsorbed water, and are not necessarily diagnostic. Frequencies followed by a question mark are of uncertain significance. Features in the spectrum approaching the lower frequency limit of $225\,\text{cm}^{-1}$ should be treated with some caution as they may be instrument artifacts due to low energy transmission.

Notes and References:
    Any information relevant to the specimen and spectrum is given here, also any polymorphism and relationships with other minerals. Mineral names in bold indicate that a spectrum of that species is included in this collection. References given have been selected to include, where possible, those involving infrared investigation, spectra or structural information. Reference may also be made to named compilations as follows:

Moenke	Moenke, H. (1962, 1966) *Mineralspektren, Parts I and II*, Akademie-Verlag, Berlin.
Nyquist and Kagel	Nyquist, A. and Kagel, R.O. (1971) *Infrared Spectra of Inorganic Compounds*, Academic Press, New York.
Farmer	Farmer, V.C. (Ed.) (1974) *The Infrared Spectra of Minerals*, Monograph no. 4, Mineralogical Society, London.
Sadtler	Ferraro, J.R. (Ed.) (1982) *Infrared Spectra Handbook of Minerals and Clays*, Sadtler Research Laboratories, Philadelphia.
Suhner	Suhner, B. (1986) *Infrarot-spektren von Mineralien, Parts 1 and 2*.

General references

Infrared Spectra of Minerals and Related Inorganic Compounds
Gadsden, J.A. (1975)
Butterworth, London.

A systematic classification of nonsilicate minerals
Ferraiolo, J.A. (1982)
Bulletin of the American Museum of Natural History, **172** (1).

The Mineral Database
Aleph Enterprises (1989)
Aleph, Livermore, California.

Encyclopedia of Minerals, 2nd Edition
Roberts, W.L., Campbell, T.J. and Rapp, G.R. (1990)
Van Nostrand Reinhold, New York.

Glossary of Mineral Species
Fleischer, M. and Mandarino, J.A. (1991)
The Mineralogical Record Inc., Tucson.

Mineral Reference Manual
Nickel, E.H. and Nichols, M.C. (1991)
Van Nostrand Reinhold, New York.

Hey's Mineral Index, 3rd Edition
Clark, A.M. (1993)
Chapman & Hall, London.

Index of spectra by mineral name

Alstonite
Alumohydrocalcite
Ancylite-(Ce)
Andersonite
Ankerite
Aragonite
Artinite
Aurichalcite
Azurite

Barentsite
Barstowite
Barytocalcite
Bastnäsite-(Ce)
Bastnäsite-(La)
Bayleyite
Benstonite
Beyerite
Bismutite
Brenkite
Brugnatellite
Burbankite

Calcite
Callaghanite
Canavesite
Carbocernaite
Carbonate-cyanotrichite
Cerussite
Coalingite
Cordylite-(Ce)

Dawsonite
Defernite
Desautelsite
Dolomite
Donnayite-(Y)
Dresserite
Dundasite
Dypingite

Gaspeite

Gaylussite
Glaukosphaerite

Harkerite
Hellyerite
Huntite
Hydrocerussite
Hydromagnesite
Hydrotalcite
Hydrozincite (2)

Ikaite
Indigirite

Kambaldaite
Kamotoite-(Y)
Kimuraite-(Y)
Kolwezite
Kutnohorite

Lanthanite-(La)
Leadhillite
Liebigite
Lokkaite

Macphersonite
Magnesite
Malachite
Manasseite (2)
Manganotychite
Mcguinnessite
Mckelveyite-(Y)
Mineevite-(Y)
Monohydrocalcite
Montroyalite

Nahcolite
Nesquehonite
Norsethite
Northupite
Nyerereite

Otavite

Paralstonite
Parisite-(Ce)
Phosgenite
Pirssonite
Pokrovskite
Pyroaurite

Rhodochrosite
Rosasite
Roubaultite

Sabinaite
Scarbroite
Schröckingerite
Sharpite
Shortite
Siderite
Sjögrenite
Smithsonite
Sphaerocobaltite
Stenonite
Strontianite
Strontiodresserite
Susannite
Synchysite-(Y)

Takovite
Trona
Tunisite
Tychite

Vaterite
Voglite

Weloganite
Witherite
Wyartite

Zaratite
Zellerite
Znucalite

Index of spectra by chemical class

Anhydrous normal carbonates

Alstonite	$BaCa(CO_3)_2$
Ankerite	$Ca(Fe,Mg,Mn)(CO_3)_2$
Aragonite	$Ca(CO_3)$
Barytocalcite	$BaCa(CO_3)_2$
Benstonite	$(Ba,Sr)_6(Ca,Mn)_6Mg(CO_3)_{13}$
Beyerite	$(Ca,Pb)Bi_2(CO_3)_2O_2$ *or* $Ca(BiO_2)(CO_3)$
Bismutite	$Bi_2(CO_3)O_2$
Burbankite	$(Na,Ca)_3(Sr,Ba,Ce)_3(CO_3)_5$
Calcite	$CaCO_3$
Carbocernaite	$(Ca,Na)(Sr,Ce,Ba)(CO_3)_2$
Cerussite	$PbCO_3$
Dolomite	$CaMg(CO_3)_2$
Gaspéite	$(Ni,Mg,Fe)CO_3$
Huntite	$CaMg_3(CO_3)_4$
Kutnohorite	$Ca(Mn,Mg,Fe)(CO_3)_2$
Magnesite	$MgCO_3$
Norsethite	$BaMg(CO_3)_2$
Nyerereite	$Na_2Ca(CO_3)_2$
Otavite	$CdCO_3$
Paralstonite	$BaCa(CO_3)_2$
Rhodochrosite	$MnCO_3$
Sabinaite	$Na_4Zr_2TiO_4(CO_3)_4$
Shortite	$Na_2Ca_2(CO_3)_3$
Siderite	$FeCO_3$
Smithsonite	$ZnCO_3$
Sphaerocobaltite	$CoCO_3$
Strontianite	$SrCO_3$
Vaterite	$CaCO_3$
Witherite	$BaCO_3$

Hydrated normal carbonates

Andersonite	$Na_2Ca(UO_2)(CO_3)_3 \cdot 6H_2O$
Bayleyite	$Mg_2(UO_2)(CO_3)_3 \cdot 18H_2O$
Donnayite-(Y)	$Sr_3NaCaY(CO_3)_6 \cdot 3H_2O$
Gaylussite	$Na_2Ca(CO_3)_2 \cdot 5H_2O$
Hellyerite	$NiCO_3 \cdot 6H_2O$

Ikaite	$CaCO_3 \cdot 6H_2O$
Kamotoite-(Y)	$Y_2O_4(UO_2)_4(CO_3)_3 \cdot 14.5H_2O$
Kimuraite-(Y)	$CaY_2(CO_3)_4 \cdot 6H_2O$
Lanthanite-(La)	$(La,Ce)_2(CO_3)_3 \cdot 8H_2O$
Liebigite	$Ca_2(UO_2)(CO_3)_3 \cdot 11H_2O$
Lokkaite	$CaY_4(CO_3)_7 \cdot 9H_2O$
Mckelveyite-(Y)	$Ba_3Na(Ca,U)Y(CO_3)_6 \cdot 3H_2O$
Monohydrocalcite	$CaCO_3 \cdot H_2O$
Pirssonite	$Na_2Ca(CO_3)_2 \cdot 2H_2O$
Voglite	$Ca_2Cu(UO_2)(CO_3)_4 \cdot 6H_2O$
Weloganite	$Sr_3Na_2Zr(CO_3)_6 \cdot 3H_2O$
Zellerite	$Ca(UO_2)(CO_3)_2 \cdot 5H_2O$

Anhydrous carbonates with hydroxyl and/or halogen

Aurichalcite	$(Zn,Cu)_5(CO_3)_2(OH)_6$
Azurite	$Cu_3(CO_3)_2(OH)_2$
Barentsite	$Na_7AlH_2(CO_3)_4F_4$
Bastnäsite-(Ce)	$(Ce,La)(CO_3)F$
Bastnäsite-(La)	$(La,Ce)(CO_3)F$
Brenkite	$Ca_2(CO_3)F_2$
Cordylite-(Ce)	$Ba(Ce,La)_2(CO_3)_3F_2$
Dawsonite	$NaAl(CO_3)(OH)_2$
Glaukosphaerite	$(Cu,Ni)_2(CO_3)(OH)_2$
Hydrocerussite	$Pb_3(CO_3)_2(OH)_2$
Hydrozincite (2)	$Zn_5(CO_3)_2(OH)_6$
Kolwezite	$(Cu,Co)_2(CO_3)(OH)_2$
Malachite	$Cu_2(CO_3)(OH)_2$
Mcguinnessite	$(Mg,Cu)_2(CO_3)(OH)_2$
Northupite	$Na_3Mg(CO_3)_2Cl$
Parisite-(Ce)	$Ca(Ce,La)_2(CO_3)_3F_2$
Phosgenite	$Pb_2(CO_3)Cl_2$
Rosasite	$(Cu,Zn)_2(CO_3)(OH)_2$
Stenonite	$(Sr,Ba,Na)_2Al(CO_3)F_5$
Synchisite-(Y)	$Ca(Y,Ce)(CO_3)_2F$
Tunisite	$NaCa_2Al_4(CO_3)_4(OH)_8Cl$

Hydrated carbonates with hydroxyl and/or halogen

Alumohydrocalcite	$CaAl_2(CO_3)_2(OH)_4 \cdot 3H_2O$
Ancylite-(Ce)	$SrCe(CO_3)_2(OH) \cdot H_2O$
Artinite	$Mg_2(CO_3)(OH)_2 \cdot 3H_2O$
Barstowite	$3PbCl_2 \cdot PbCO_2 \cdot H_2O$
Brugnatellite	$Mg_6Fe(CO_3)(OH)_{13} \cdot 4H_2O$
Callaghanite	$Cu_2Mg_2(CO_3)(OH)_6 \cdot 2H_2O$
Coalingite	$Mg_{10}Fe_2(CO_3)(OH)_{24} \cdot 2H_2O$
Defernite	$Ca_3(CO_3)(OH,Cl)_4 \cdot H_2O$
Desautelsite	$Mg_6Mn_2(CO_3)(OH)_{16} \cdot 4H_2O$
Dresserite	$Ba_2Al_4(CO_3)_4(OH)_8 \cdot 3H_2O$
Dundasite	$PbAl_2(CO_3)_2(OH)_4 \cdot H_2O$
Dypingite	$Mg_5(CO_3)_4(OH)_2 \cdot 5H_2O$
Hydromagnesite	$Mg_5(CO_3)_4(OH)_2 \cdot 4H_2O$
Hydrotalcite	$Mg_6Al_2(CO_3)(OH)_{16} \cdot 4H_2O$

Indigirite	$Mg_2Al_2(CO_3)_4(OH)_2 \cdot 15H_2O$
Kambaldaite	$NaNi_4(CO_3)_3(OH)_3 \cdot 3H_2O$
Manasseite (2)	$Mg_6Al_2(CO_3)(OH)_{16} \cdot 4H_2O$
Montroyalite	$Sr_4Al_8(CO_3)_3[(OH),F]_{26} \cdot 10-11H_2O$
Pokrovskite	$Mg_2(CO_3)(OH)_2 \cdot 0 \cdot 5H_2O$
Pyroaurite	$Mg_6Fe_2(CO_3)(OH)_{16} \cdot 4H_2O$
Roubaultite	$Cu_2O_2(UO_2)_3(CO_3)_2(OH)_2 \cdot 4H_2O$
Scarbroite	$Al_5(OH)_{13}(CO_3) \cdot 5H_2O$
Sharpite	$Ca(UO_2)_6(CO_3)_5(OH)_4 \cdot 6H_2O$
Sjögrenite	$Mg_6Fe_2(CO_3)(OH)_{16} \cdot 4H_2O$
Strontiodresserite	$(Sr,Ca)Al_2(CO_3)_2(OH)_4 \cdot H_2O$
Takovite	$Ni_6Al_2[(CO_3),(OH)](OH)_{16} \cdot 4H_2O$
Wyartite	$Ca_3U(UO_2)_6(CO_3)_2(OH)_{18} \cdot 3-5H_2O$
Zaratite	$Ni_3(CO_3)(OH)_4 \cdot 4H_2O$
Znucalite	$Zn_{12}Ca(UO_2)(CO_3)_3(OH)_{22} \cdot 4H_2O$

Acid carbonates

Nahcolite	$NaHCO_3$
Nesquehonite	$Mg(HCO_3)(OH) \cdot 2H_2O$
Trona	$Na_3(CO_3)(HCO_3) \cdot 2H_2O$

Compound carbonates

Canavesite	$Mg_2(CO_3)(HBO_3) \cdot 5H_2O$
Carbonate-cyanotrichite	$Cu_4Al_2(CO_3,SO_4)(OH)_{12} \cdot 2H_2O$
Harkerite	$Ca_{24}Mg_8Al_2Si_8[O,(OH)]_{32}(BO_3)_8(CO_3)_8(H_2O,Cl)$
Leadhillite	$Pb_4(SO_4)(CO_3)_2(OH)_2$
Macphersonite	$Pb_4(SO_4)(CO_3)_2(OH)_2$
Manganotychite	$Na_6(Mn,Fe,Mg)_2(CO_3)_4(SO_4)$
Mineevite-(Y)	$Na_{25}BaY_2(CO_3)_{11}(HCO_3)_4(SO_4)_2F_2Cl$
Schröckingerite	$NaCa_3(UO_2)(CO_3)_3(SO_4)F \cdot 10H_2O$
Susannite	$Pb_4(SO_4)(CO_3)_2(OH)$
Tychite	$Na_6Mg_2(SO_4)(CO_3)_4$

The spectra

ALSTONITE

Formula:	**BaCa(CO₃)₂**
Chemical class:	**Anhydrous normal carbonate**
Chemical type:	**AB(XO₃)₂**

Crystal system: **Triclinic, pseudo orthorhombic**
Mineral group: **Aragonite**
Space group: **P1**

Specimen:	**BM 41707 Large, colourless, prismatic crystals with witherite and calcite.**
Source:	**Brownley Hill mine, Nenthead, Alston, Cumbria, U.K. (Type locality).**
Spectrum ref. no.:	**IR2666**
Sample medium:	**KBr disk**
XRD:	**7451F (std)**
Composition:	**Ba:Ca:Sr = 1:1·0·2**

Notes

Trimorphous with **barytocalcite** and **paralstonite**.
The spectrum differs from that of paralstonite only in the 700 cm⁻¹ region and is similar to, but distinguishable from, that of barytocalcite.
Compare spectrum with those of other members of **aragonite** group.

References:

1. Kaushansky P. & Yariv S. (1986) The interactions between calcite particles and aqueous solutions of magnesium, barium or zinc chlorides.
   Applied Geochemistry, **1**(5), pp.607-618.

2. Scheetz B.E. & White W.B. (1977) Vibrational spectra of the alkaline earth double carbonates.
   American Mineralogist, **62**(12), pp.36-50.

3. Rossman G.R. & Squires R.L. (1974) The occurrence of alstonite at Cave-in-Rock, Illinois.
   Mineralogical Record, **5**(6), pp.266-269.

Peak Table cm⁻¹

[3413]	800
2923	752
2856	725?
2493	**708**
2466	**701**
1766	**691**
1754	518
1734	**464**
1503	**302**
1458	251
1436	
1413	
1390	
1171	
1086	
1063	
894	
861	
855	

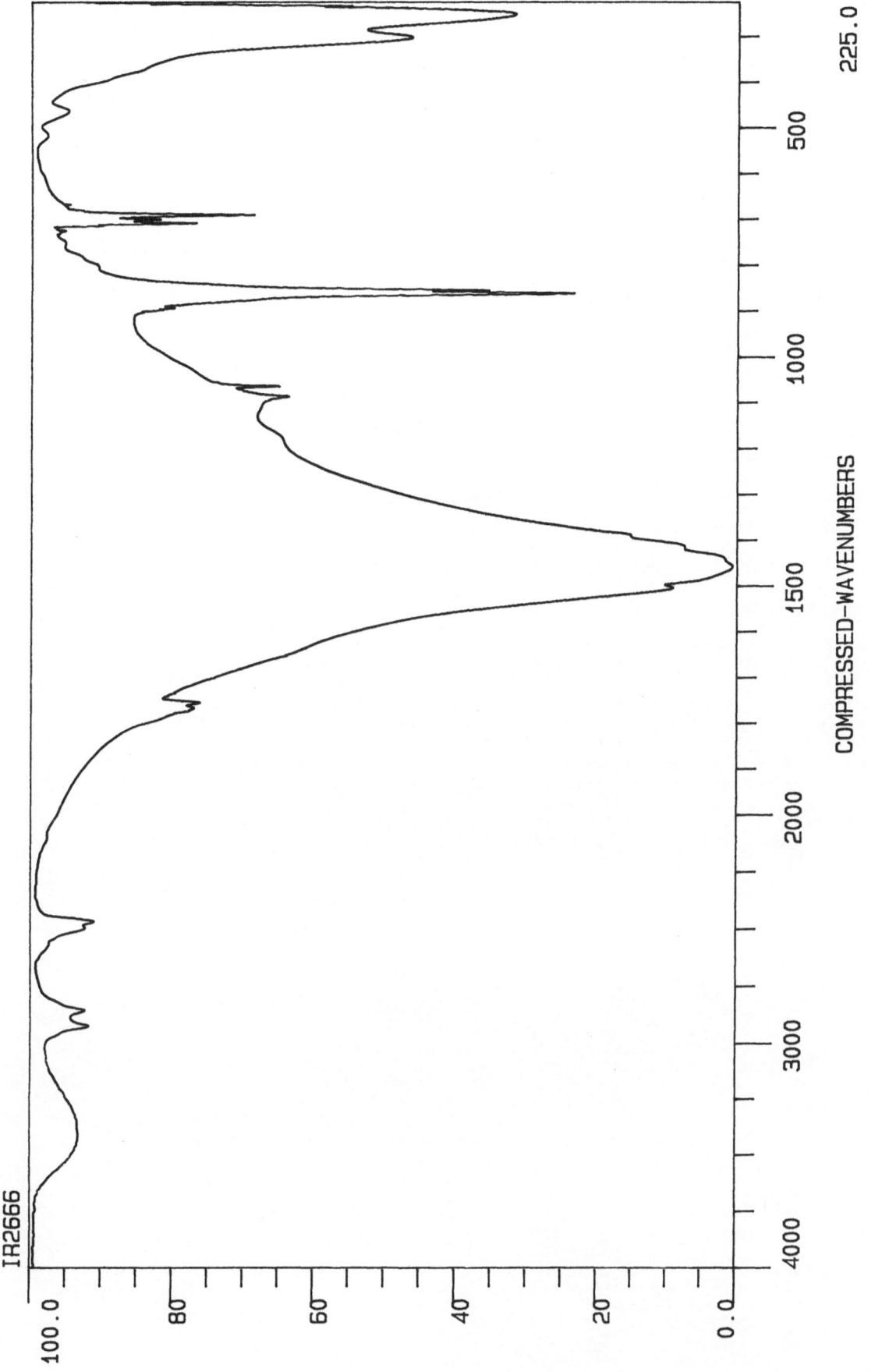

IR2666

% TRANSMITTANCE

COMPRESSED-WAVENUMBERS

ALSTONITE

IR2666

% TRANSMITTANCE

WAVENUMBERS

ALSTONITE [expanded detail]

ALUMOHYDROCALCITE

Formula:	$CaAl_2(CO_3)_2(OH)_4 \cdot 3H_2O$
Chemical class:	Hydrated carbonate with hydroxyl or halogen
Chemical type:	$A_mB_n(XO_3)_pZ_q \cdot xH_2O$
Crystal system:	Triclinic ?
Mineral group:	
Space group:	?

Specimen:	BM 1937,1377 Lavender, silky, fibrous, radiating with calcite on serpentine.
Source:	Ruben mine, Nowa Ruda, Poland.
Spectrum ref. No.:	IR2796
Sample medium:	KBr disk
XRD:	7906F (std)
Composition:	Ca:Al = 1:1·8 + minor Si

Notes

The spectrum is more complex i.e. better resolved, than that given in ref. 3, but is otherwise identical.

References:

1. Ryback G. (1988)
   Alumohydrocalcite from Scarborough, North Yorkshire, and Weston Favell, Northamptonshire.
   Journal of the Russell Society, **2**(1), pp.9-12.

2. Srebrodol'skiy B.I. (1976)
   Alumohydrocalcites.
   International Geological Revue, **18** (3), pp.321-328.

3. Kautz K. (1969)
   Electron microscope and infrared investigation of alumohydrocalcite.
   Neues Jahrbuch für Mineralogie, Monatsheft, No.3, pp.130-137
   (German with English summary).

Peak Table cm^{-1}

3702	1031
3653	1007
3622	967
3364	867
3145	798
2961	730
2923	661
2853	570
2516	523
2187	472
1995	426
1834	343
1798	250?
1673	
1519	
1420	
1399	
1113	
1099	

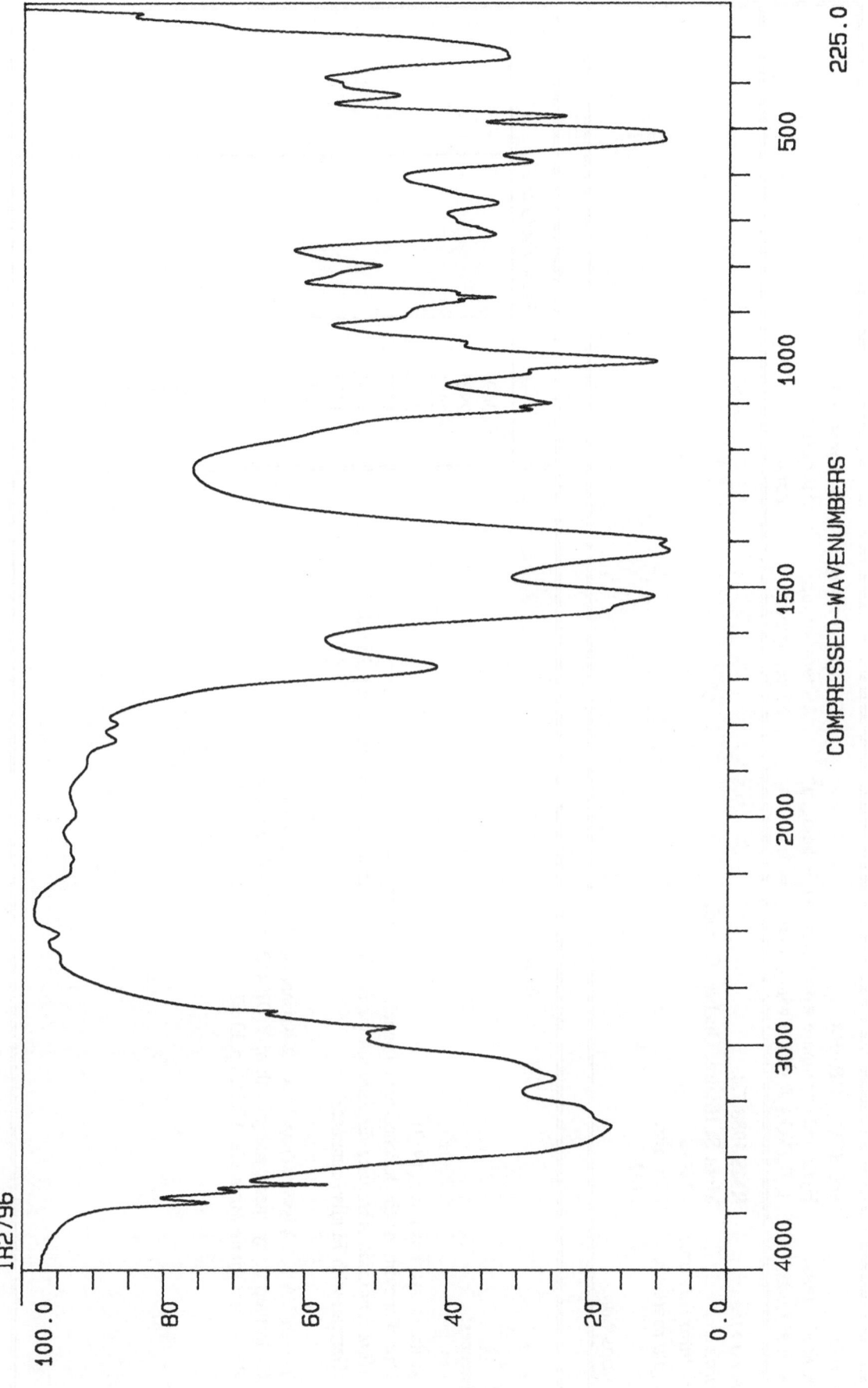

IR2796

% TRANSMITTANCE

100.0   80   60   40   20   0.0

4000   3000   2000   1500   1000   500   225.0

COMPRESSED-WAVENUMBERS

ALUMOHYDROCALCITE

ANCYLITE-(Ce)

Formula:	$SrCe(CO_3)_2(OH) \cdot H_2O$
Chemical class:	Hydrated carbonate with hydroxyl or halogen
Chemical type:	$A_m B_n (XO_3)_p Z_q \cdot xH_2O$ with (m+n):p = 1:1
Specimen:	RMS 1988.7.8.
Source :	Mont St Hilaire, Quebec, Canada.
Spectrum ref. no.:	IR2916
Sample medium:	KBr disk
XRD:	4142
Composition:	

Crystal system:	Orthorhombic
Mineral group:	Ancylite
Space group:	Pmcn

Peak Table cm⁻¹

3494
2925
2856
2532
1775
1455
1373
1074
858
771
727
715
703
695
294

Notes

References:

1. Walter F. & Postl W. (1983)
   Calcio ancylite of the Kalcherkogel Tunnel, Pack, Styria.
   Mitteilungsblatt Abteilung für Mineralogie am Landesmuseum Joanneum, **51**, pp.25-28.
   (German with English summary).

2. Tareen J.A.K., Viswanathiah M.N. & Krishnamurthy K.V. (1980)
   Hydrothermal synthesis and growth of Y(OH)CO₃ ancylite-like phases.
   Revue de Chimie Minerale, **17**(1), pp.50-57.

IR2916

% TRANSMITTANCE

COMPRESSED-WAVENUMBERS

ANCYLITE- (Ce)

ANDERSONITE

Formula:	Na$_2$Ca(UO$_2$)(CO$_3$)$_3$·6H$_2$O
Chemical class:	Hydrated normal carbonate
Chemical type:	A$_m$B$_n$(XO$_3$)$_p$·xH$_2$O where (m+n):p > 1:1

Crystal system:	Trigonal
Mineral group:	Rutherfordine
Space group:	R$\bar{3}$

Specimen:	BM 1967,267 Yellow/green transparent isolated crystals.
Source:	Atomic King No.2 mine, Cane Wash, San Juan Co., Utah, U.S.A.
Spectrum ref. no.:	IR2859
Sample medium:	KBr disk
XRD:	8099F matches PDF 20-1092 andersonite.
Composition:	Na:Ca:U = 0·7:1:2 + trace Zn & Si

Peak Table cm⁻¹

3548	852
3413	**847**
3216	764
2903	**728**
2602	**700**
2418	671
2360	632
2330	**542**
2082	**476**
1816	**425**
1659	343
1575	291
1525	
1382	
1093	
1080	
954	
914	
903	

Notes

The spectrum has a strong peak at 3548 cm⁻¹ which indicates the presence of an (OH) group not shown in the formula. Matches Suhner (5-35 A) andersonite, except for the lack of a peak at 1020 cm⁻¹. Semi-quantitative analysis of this specimen gave a lower sodium content than indicated by the formula.

References:

1. Cejka J., Urbanec Z. & Cejka J Jr. (1987)
   Contribution to the crystal chemistry of andersonite.
   Neues Jahrbuch für Mineralogie, Monatshefte, (11), pp.488-501.

2. Urbanec Z. & Cejka J. (1979)
   Infrared spectra of liebigite, andersonite, voglite, and schroeckingerite.
   Collection of Czechoslovak Chemical Communications, **44**(1), pp.10-23.

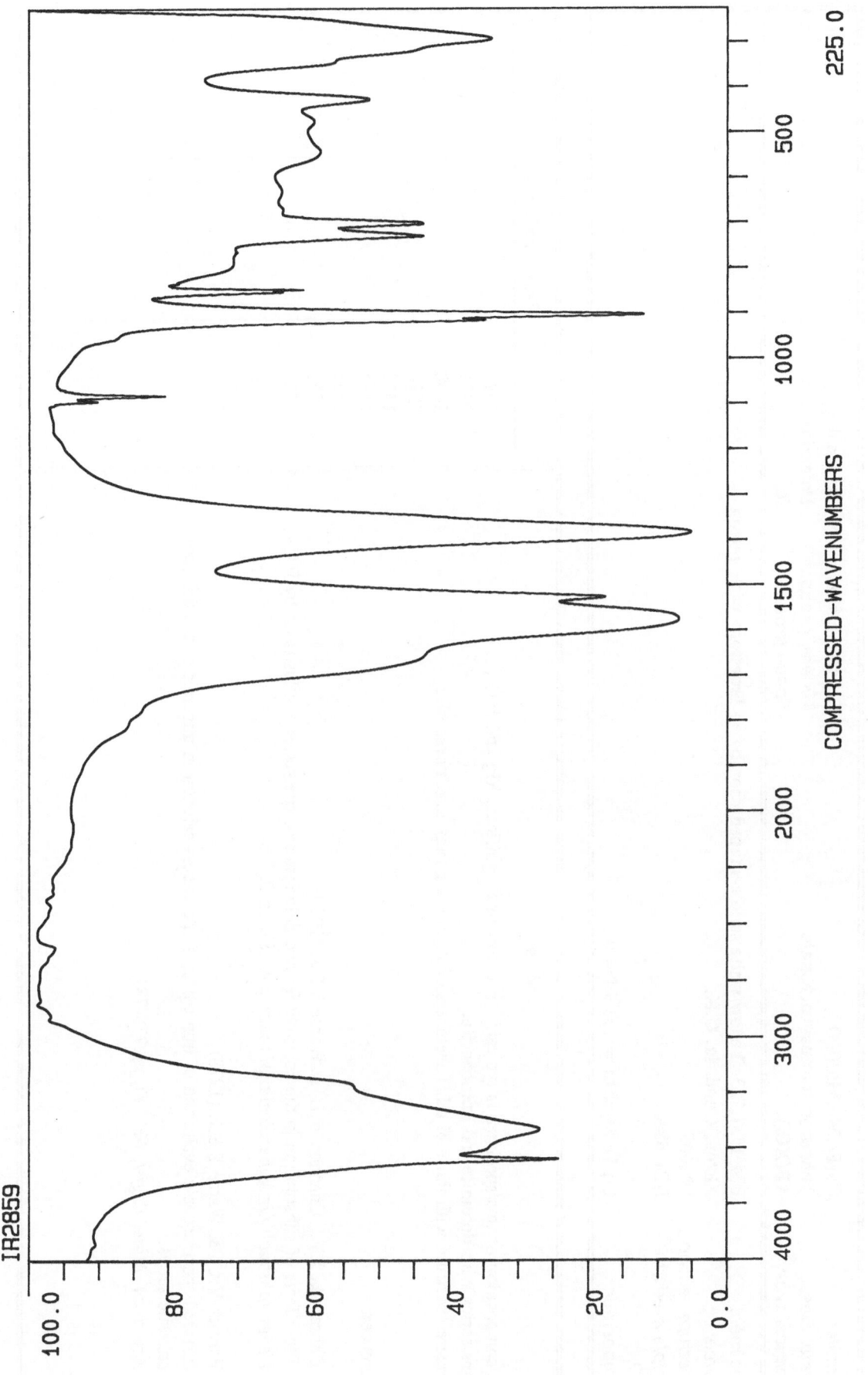

IR2859

% TRANSMITTANCE

100.0 80 60 40 20 0.0

COMPRESSED-WAVENUMBERS

4000 3000 2000 1500 1000 500 225.0

ANDERSONITE

ANKERITE

Formula:	**Ca(Fe,Mg,Mn)(CO₃)₂**
Crystal class:	**Anhydrous normal carbonate**
Chemical type:	**AB(XO₃)₂**

Crystal system: **Trigonal**
Mineral group: **Dolomite**
Space group: **R3̄**

Specimen:	**BM 1921,723** Yellow/white saddle-shaped rhombs on limestone with galena.
Source:	**Alston, Cumbria, U.K.**
Spectrum ref. no.:	**IR2665**
Sample medium:	**KBr disk**
XRD:	
Composition:	**Ca:Fe:Mg:Mn = 1:0·5:0·4:0·1**

Notes

The composition of this specimen is typical, i.e. it contains significant Mg and Mn.
Forms series with **dolomite** and **kutnohorite.**
Compare spectrum with those of other members of dolomite group, e.g. **norsethite**

References:

1. Dubrawski J.V., Channon A.L. & Warne S.S.J. (1989)
   The effects of substitution in the dolomite ferroan dolomite ankerite series as illustrated by FTIR.
   Neues Jahrbuch für Mineralogie. Monatshefie, **(8)**, pp.337-344.

2. Farmer V.C. & Warne S.S.J. (1978)
   Infrared spectroscopic evaluation of iron contents and excess calcium in minerals of the dolomite ankerite series.
   American Mineralogist, **63**(7,8), pp.779-781.

Peak Table cm⁻¹

2987
2871
2608
2510
1810
1424
1091
875
725
353
324

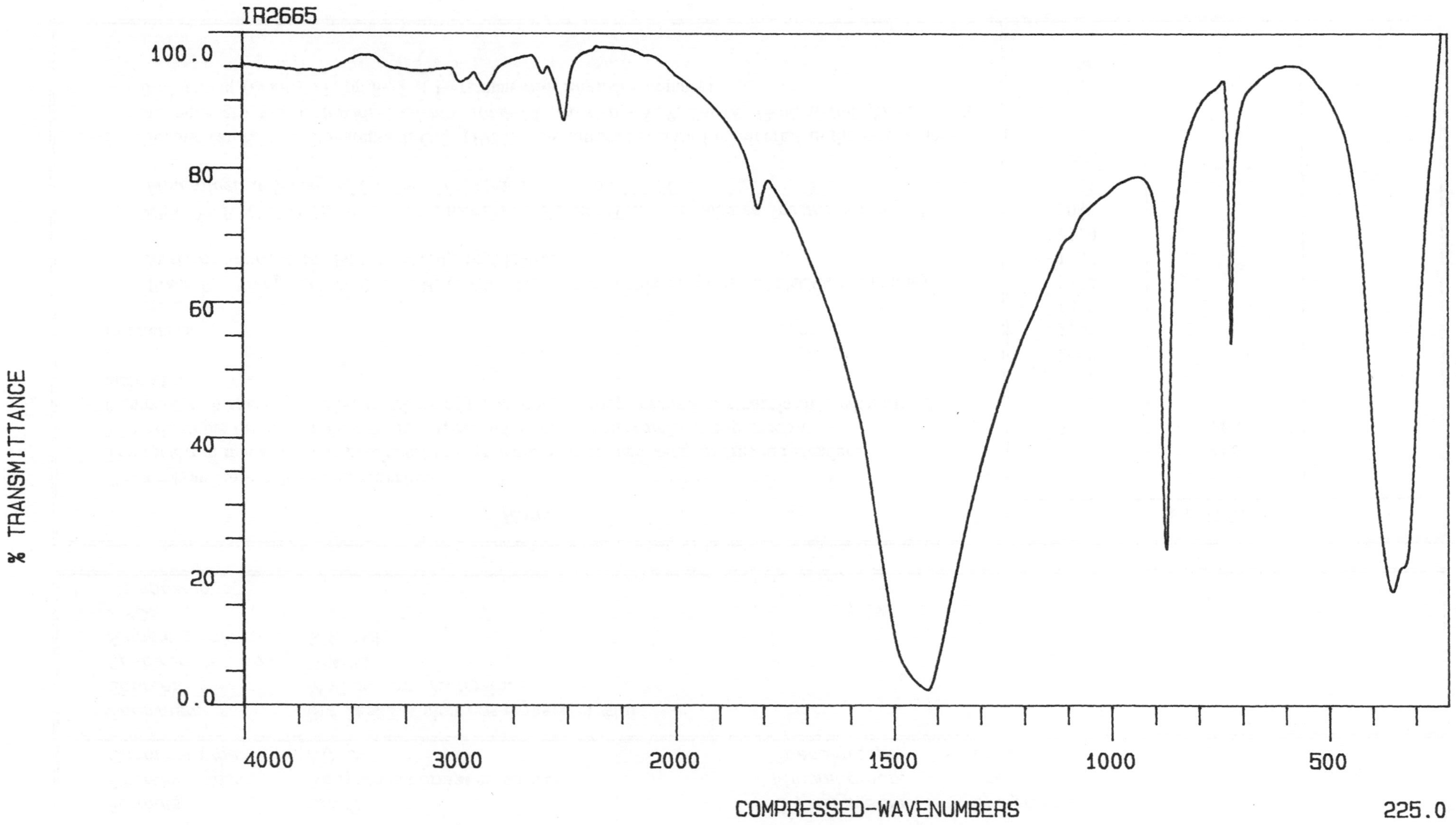

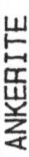

IR2665

% TRANSMITTANCE

100.0   80   60   40   20   0.0

4000   3000   2000   1500   1000   500   225.0

COMPRESSED-WAVENUMBERS

ANKERITE

ARAGONITE

Crystal system:	Orthorhombic
Mineral group:	Aragonite
Space group:	Pmcn

Formula:	$CaCO_3$
Chemical class:	Anhydrous normal carbonate.
Chemical type:	$A(XO_3)$

Specimen:	BM 26522. Colourless twinned crystals.
Source:	Molina, Aragon, Spain.
Spectrum ref. no.:	IR2602
Sample medium:	KBr disk
XRD:	
Composition:	

Peak Table cm^{-1}

[3307]	536
2921	469
2854	**268**
2546	
2522	
2499	
1789	
1477	
1384	
1167	
1119	
1083	
1034	
909	
858	
844	
713	
700	

Notes

Trimorphous with **calcite and vaterite.**
The spectrum matches those obtained from synthetic material and x-ray diffraction standard.
Typical simple orthorhombic carbonate spectrum, distinct from **calcite** group spectra.
Compare with spectra of other members of the aragonite group; **cerussite, strontianite, witherite & alstonite.**

References:

1. Frech R., Wang E.C. & Bates J.B. (1980) The I.R. and Raman spectra of $CaCO_3$ (aragonite). *Spectrochimica Acta*, Part A, **36**(10), pp.915-919.

2. White W.B. (1974) The carbonate minerals. *In:* Farmer (Ed.) *The Infrared Spectra of Minerals. Mineralogical Society of London, Monograph No. 4*, pp.227-284.

3. Gevork'yan S.V. & Povarennikh O.S. (1983) New infrared spectra for minerals in the calcite and aragonite groups. *Dopovidi Akademiyi Nauk Ukrayins'koyi RSR, Ser. B: Geologichni, Khimichni ta Biologichni Nauki.*, **11**, pp.8-12. (Ukrainian with English summary).

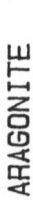

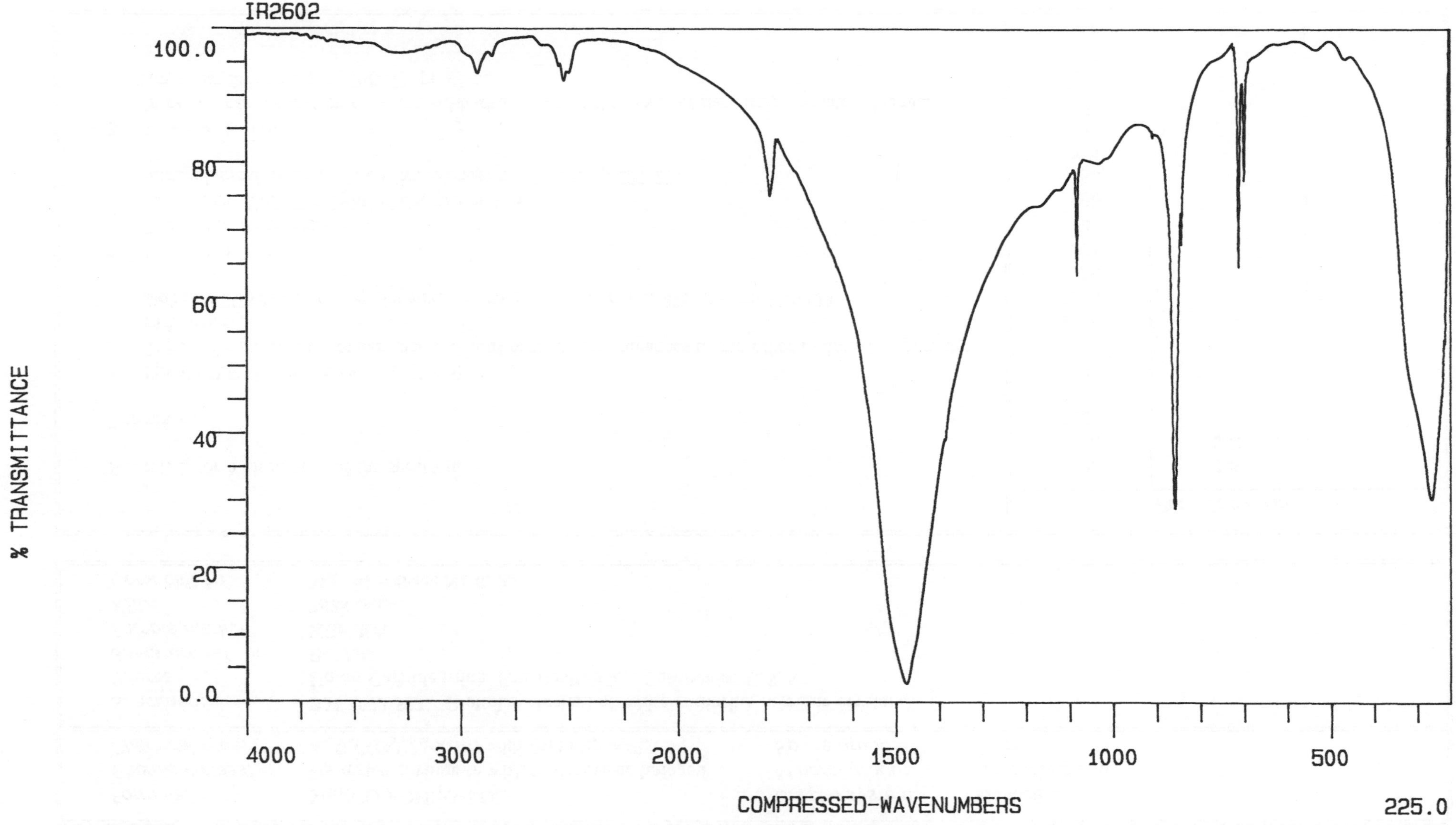
IR2602
100.0
80
60
40
20
0.0
% TRANSMITTANCE
4000
3000
2000
1500
1000
500
COMPRESSED-WAVENUMBERS
225.0

ARAGONITE

ARTINITE

Formula:	$Mg_2(CO_3)(OH)_2 \cdot 3H_2O$
Chemical class:	Hydrated carbonate with hydroxyl or halogen
Chemical type:	$A_mB_n(XO_3)_pZ_q \cdot xH_2O$ with (m+n):p = 2:1
Crystal system:	Monoclinic
Mineral group:	Hydromagnesite
Space group:	C2/m

Specimen:	BM 1973,503. Globular clusters of white radiating acicular crystals.
Source:	Union Carbide mine, San Benito Co., California, U.S.A.
Spectrum ref. no.:	IR2768
Sample medium:	KBr disk
XRD:	7828 (std)
Composition:	Mg with trace Na & Al

Peak Table cm^{-1}

3604	349
3014	281
2421	
2241	
1792	
1589	
1452	
1369	
1328	
1094	
942	
900	
850	
767	
731	
676	
514	
439	
396	

Notes

See ref. 2 for a discussion of the spectrum.

References:

1. Smolin P.P. & Ziborova T.A. (1976)
    Types of water, stoichiometry and relations between hydromagnesite and other hydrated magnesium carbonates.
    Doklady USSR Academy of Sciences, Earth Sciences Section, **226**(16), pp.130-133.

2. White W.B. (1974)
    The carbonate minerals.
    In: Farmer (Ed.) *The Infrared Spectra of Minerals,*
    Mineralogical Society of London, Monograph No. 4, pp.227-284.

3. White W.B. (1971)
    Infrared characterization of water and hydroxyl ion in the basic magnesium carbonate minerals.
    American Mineralogist, **56**(1,2), pp.46-53.

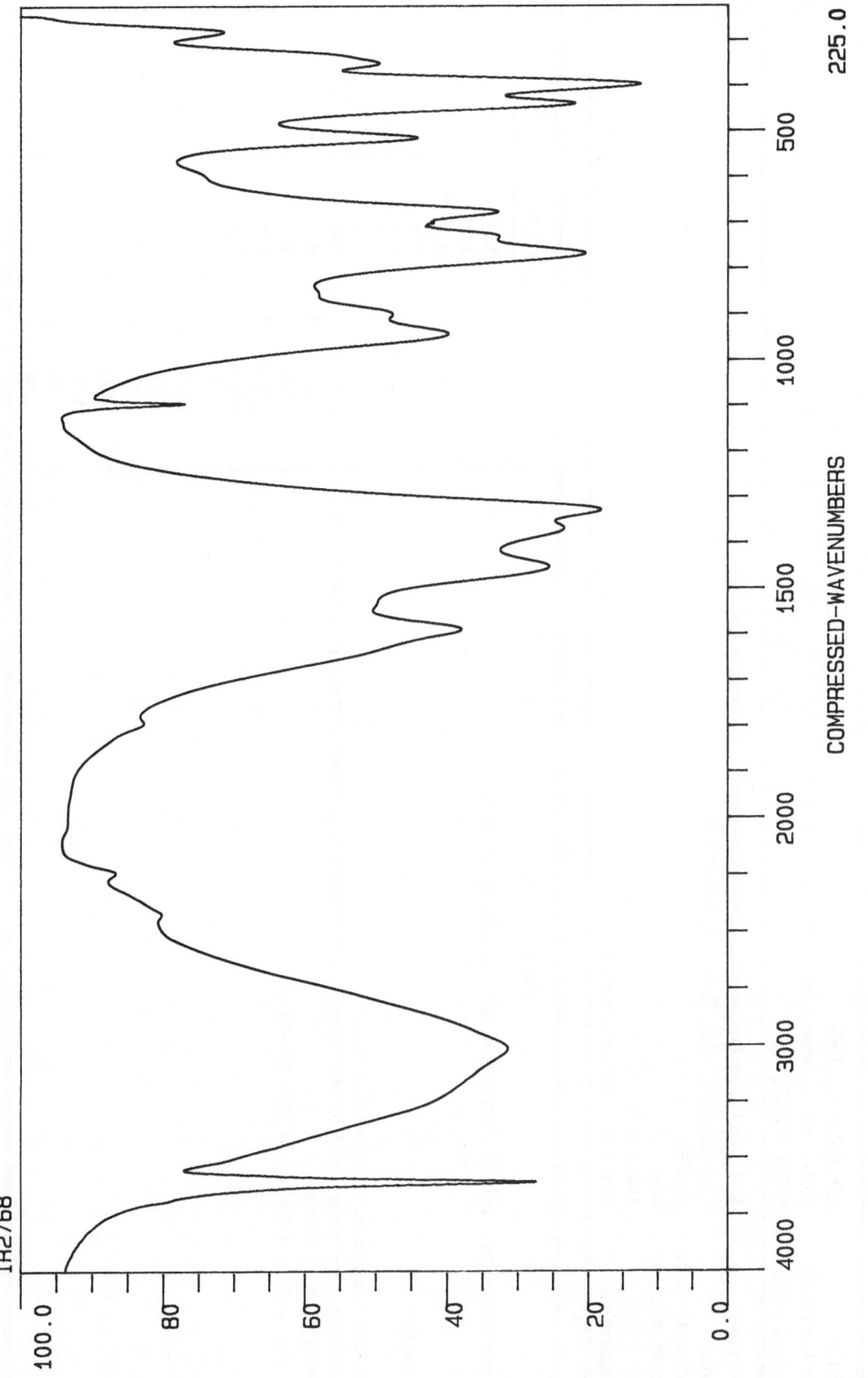

IR2768

% TRANSMITTANCE

100.0   80   60   40   20   0.0

COMPRESSED–WAVENUMBERS

4000   3000   2000   1500   1000   500   225.0

ARTINITE

AURICHALCITE

Formula:	$(Zn,Cu)_5(CO_3)_2(OH)_6$
Chemical class:	Anhydrous carbonate with hydroxyl or halogen
Chemical type:	$(AB)_5(XO_3)_2Z_q$

Crystal system:	Orthorhombic
Mineral group:	Aurichalcite
Space group:	$B22_12$

Specimen:	BM 56865 Pale blue/green spheroidal aggregates with calcite, azurite etc.
Source:	Copper Queen mine, Bisbee, Arizona, U.S.A.
Spectrum ref. no.:	IR2740
Sample medium:	KBr disk
XRD:	5989F (std)
Composition:	Zn:Cu = 1·8:1 with trace Si

Peak Table cm⁻¹

3348	841
2920	832
2682	765
2555	758
2433	742
2325	712
2117	508
1817	471
1773	411
1559	377
1505	315
1413	
1365	
1203	
1087	
1070	
1031	
979	
869	

Notes

The spectrum is quite different from that of the chemically similar rosasite.

References:

1. Braithwaite R.S.W. & Ryback G. (1962)
   Rosasite, aurichalcite, and associated minerals from Heights of Abraham, Matlock Bath, Derbyshire, with a note on infra-red spectra.
   Mineralogical Magazine, **33**(261), pp.441-449.

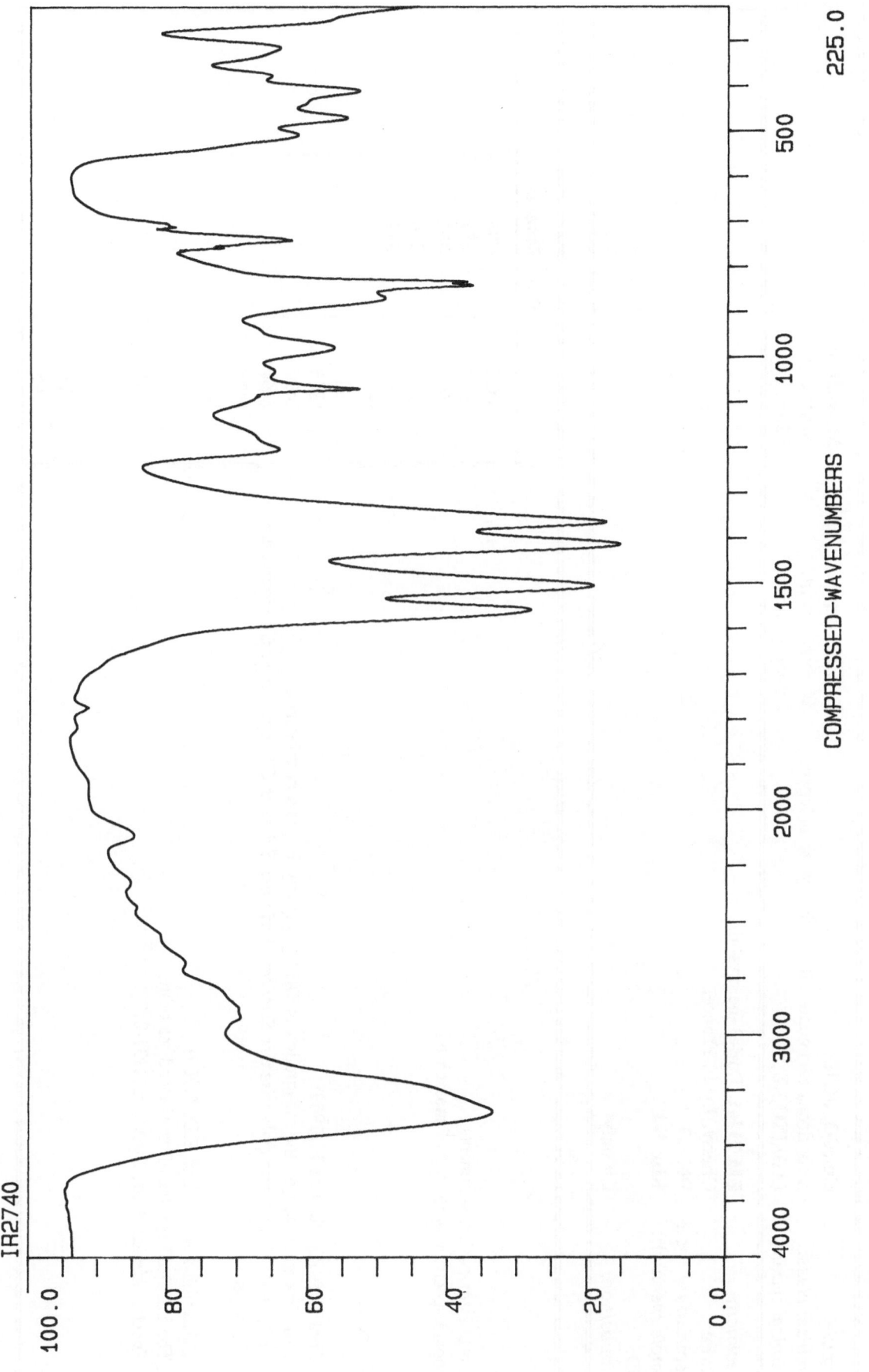

IR2740

% TRANSMITTANCE

COMPRESSED-WAVENUMBERS

AURICHALCITE

AZURITE

Formula:	$Cu_3(CO_3)_2(OH)_2$
Chemical class:	Anhydrous carbonate with hydroxyl or halogen
Chemical type:	$(AB)_3(XO_3)_2Z_q$

Crystal system:	Monoclinic
Mineral group:	Azurite
Space group:	$P2_1/c$

Specimen:	BM 91461. Dark blue crystals.
Source:	Chessy, Lyon, France.
Spectrum ref. no.:	IR2733
Sample medium:	KBr disk
XRD:	
Composition:	Cu only

Peak Table cm^{-1}

3426	496
2923	457
2593	404
2551	346
2498	312
1881	
1861	
1835	
1497	
1464	
1418	
1385	
1172	
1093	
954	
838	
820	
771	
746	

Notes

See ref. 2 for peak assignments.
Compare spectrum with that of malachite.

References:

1. Fijal J. & Zietkiewicz J. (1969)
   Experimental study on the substitution of OH^{-1} groups by F^{-1} ions in minerals.
   Bulletin de l'Académie Polonaise des Sciences. Série des Sciences Géologiques et Géographiques.
   17(1), pp.7-12.

2. Goldsmith J.A. & Ross S.D. (1969)
   The infra-red spectra of azurite and malachite.
   Spectrochimica Acta, **24**(A), pp.2131-2137.

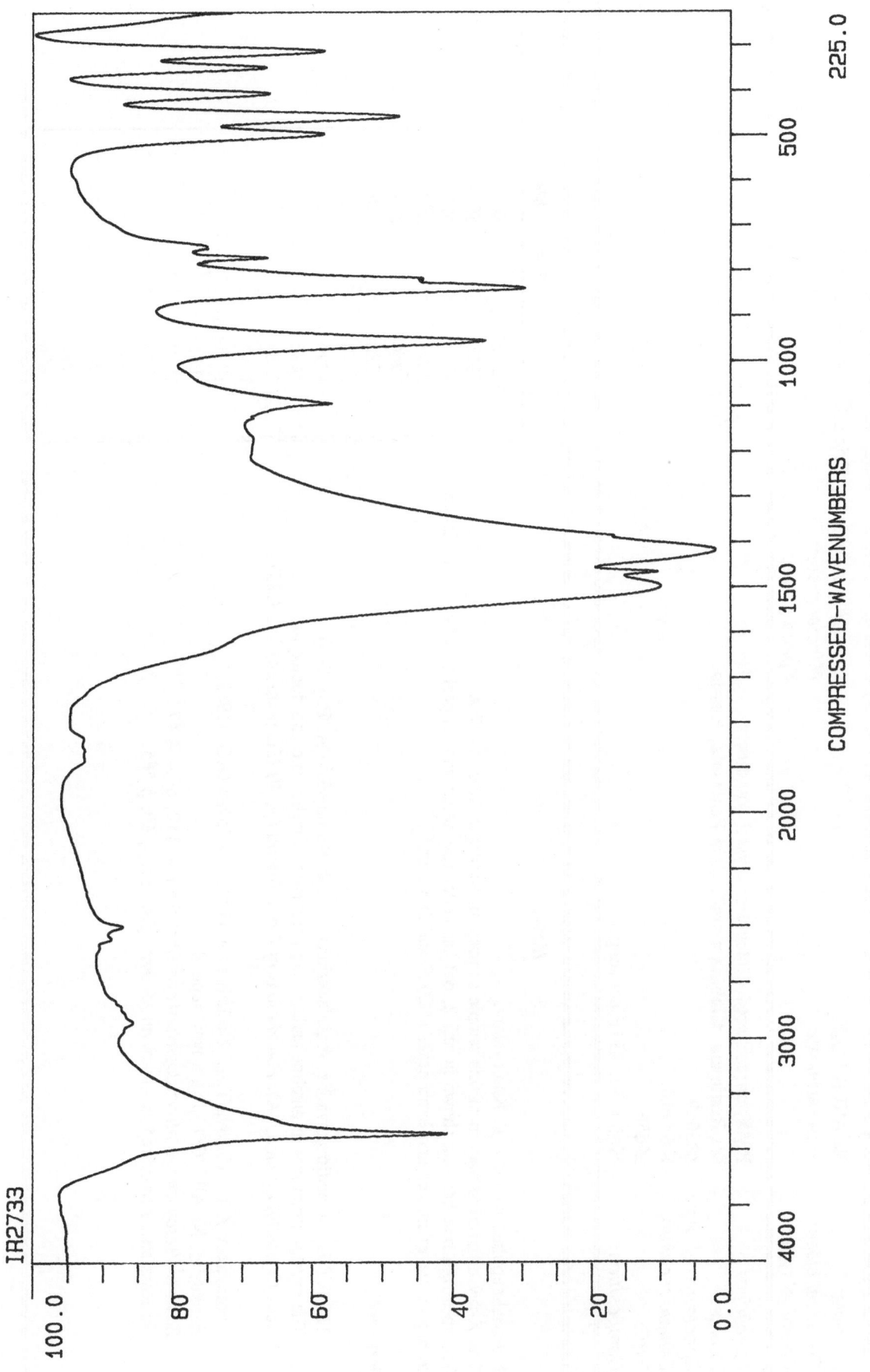

IR2733

% TRANSMITTANCE

100.0   80   60   40   20   0.0

4000   3000   2000   1500   1000   500   225.0

COMPRESSED-WAVENUMBERS

AZURITE

BARENTSITE

Formula:	$Na_7AlH_2(CO_3)_4F_4$
Chemical class:	**Acid carbonate**
Chemical type:	
Crystal system:	**Triclinic, pseudohexagonal**
Mineral group:	
Space group:	**?**

Specimen:	RMS, unregistered. Colourless crystal fragments.
Source:	Mt Restinyon, Khibina massif, Kola Peninsula, Russia.
Spectrum ref. no.:	IR3069
Sample medium:	KBr disk
XRD:	9065F
Composition:	Major Na, Al (F not sought)

Peak Table cm^{-1}

[3393]	661
2923	599
2853	429
2585	340
1966	289
1788	268
1660	
1499	
1438	
1384	
1348	
1081	
1035	
1010	
871	
839	
743	
711	
686	

Notes

Material supplied by Dr A.P. Khomyakov.

The X-ray diffraction pattern of this sample showed an additional line at 8·7 Å.

The spectrum matches that shown in ref. 2, but has better resolution and all peaks shifted by 15-20 cm^{-1} to higher wavenumbers and extra peaks at 2923 and 2833 cm^{-1} .

References:

1. Thi T.T.L., Pobedimskaya Ye. A., Nadezhina T.N. & Khomyakov A. P. (1984)
    The crystal structures of alkaline carbonates; barentsite; bonshtedtite and donnayite.
    Acta Crystallographica, (A): *Foundations of Crystallography*, 40 (Supplement), p. C257.

2. Khomyakov A.P., Kurova T.A., Nechelyustov G.N. & Piloyan G.O. (1983)
    Barentsite, $Na_7AlH_2(CO_3)_4F_4$, a new mineral.
    Zapiski Vsesoyuznogo Mineralogicheskogo Obshchestva, 112, pp.474-479.
    (in Russian). Abstracted in *American Mineralogist*, 1984, 69, p.565.

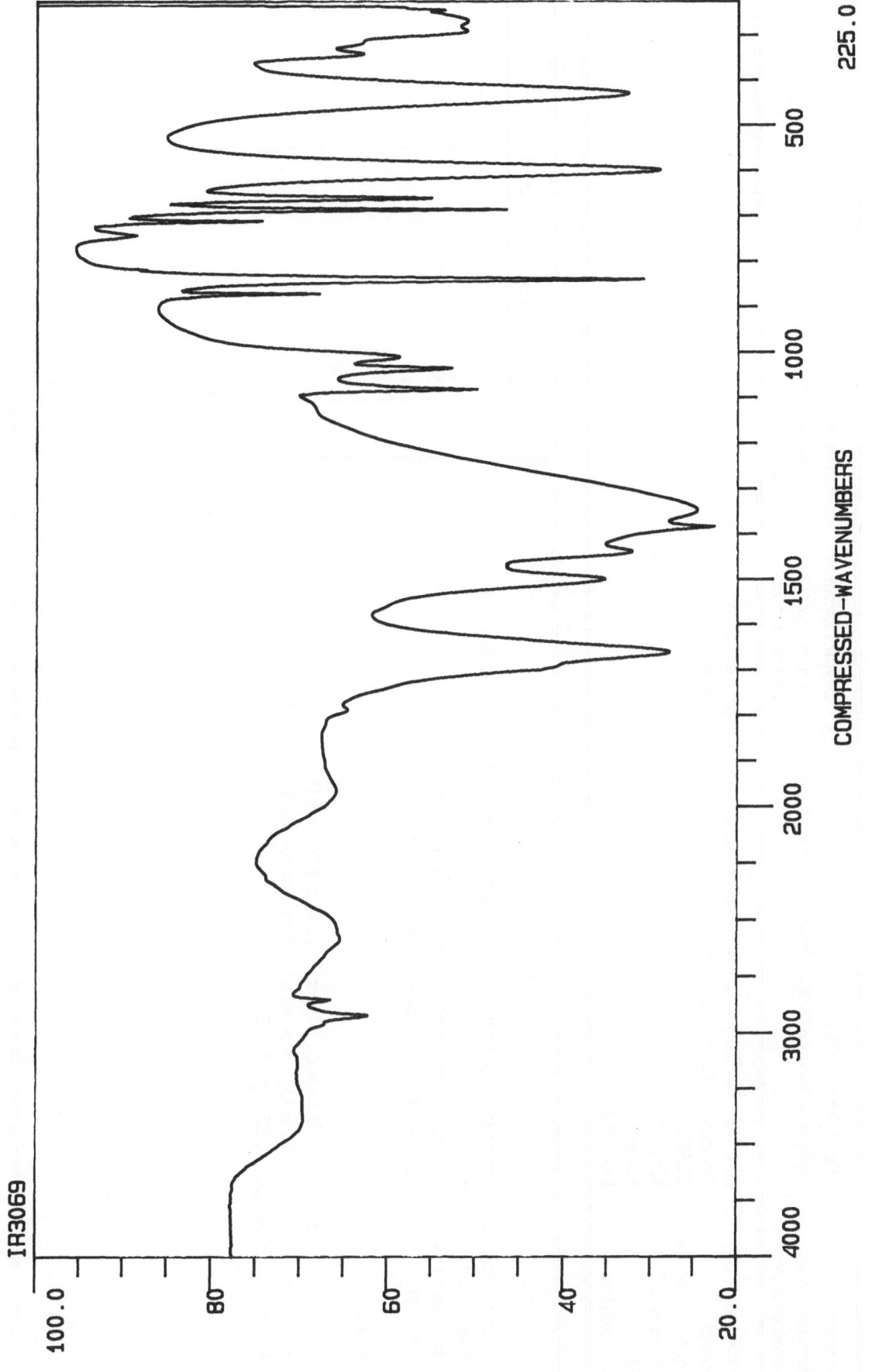

IR3069

% TRANSMITTANCE

100.0   80   60   40   20.0

4000   3000   2000   1500   1000   500   225.0

COMPRESSED-WAVENUMBERS

BARENTSITE

BARSTOWITE

Crystal system:	Monoclinic
Mineral group:	Phosgenite ?
Space group:	$P2_1/m$

Formula:	$3PbCl_2 \cdot PbCO_3 \cdot H_2O$
Chemical class:	Hydrated carbonate with hydroxyl or halogen
Chemical type:	
Specimen:	BM 1990,25 Tiny colourless crystal aggregates. (Type specimen).
Source:	Bounds Cliff, St Endellion, Cornwall, U.K. (Type locality).
Spectrum ref. no.:	IR3072
Sample medium:	KBr disk
XRD:	See ref 1.
Composition:	See ref.1.

Notes

A full description of this material is given in ref.1, including comparison of the spectrum with those of phosgenite and cerussite.

The spectrum is similar to, but distinguishable from that of phosgenite.

The small peak at 1385 cm^{-1} may be due to impurity in the KBr medium.

References:

1. Stanley C.J., Jones G.C., Hart A.D., Keller P. & Lloyd D. (1991)
   Barstowite, $3PbCl_2 \cdot PbCO_3 \cdot H_2O$, a new mineral from Bounds Cliff, St Endellion, Cornwall.
   Mineralogical Magazine, 55, pp.121-125.

Peak Table cm^{-1}

3399
2923
2854
1768
1716
1619
1438
1385?
1339
1106
1051
845
719
671
598
467
394
268

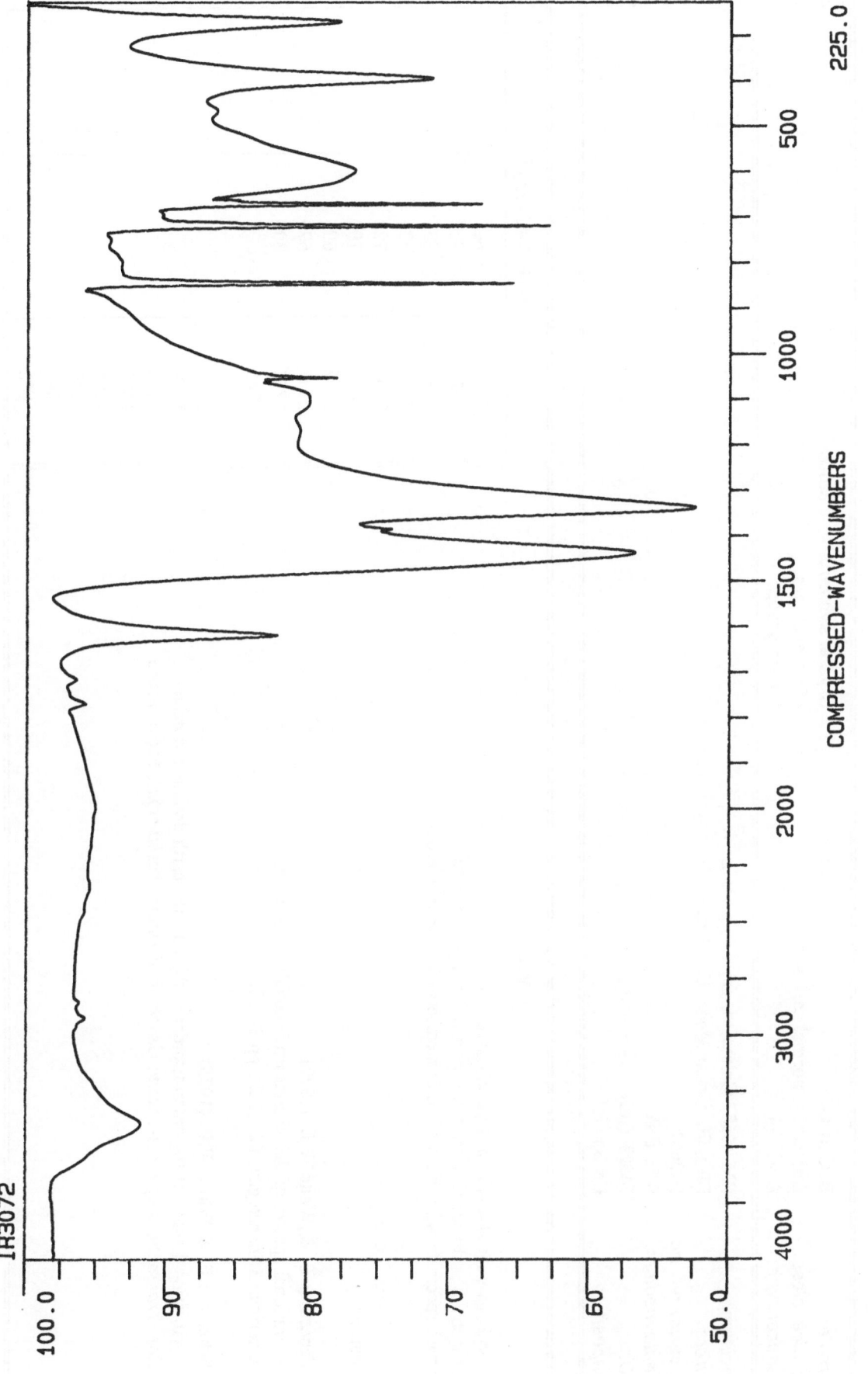

IR3072

% TRANSMITTANCE

COMPRESSED-WAVENUMBERS

BARSTOWITE

BARYTOCALCITE

Formula:	BaCa(CO₃)₂

Formula: $BaCa(CO_3)_2$
Chemical class: Anhydrous normal carbonate
Chemical type: $AB(XO_3)_2$

Crystal system: Monoclinic
Mineral group: Aragonite
Space group: $P2_1/m$

Specimen: BM 40687 Colourless crystals on massive.
Source: Bleagill, Alston Moor, Cumbria, U.K. (Type locality).
Spectrum ref. no.: IR2667
Sample medium: KBr disk
XRD: 7452F (std)
Composition: Ca:Ba 1:1

Notes

Trimorphous with alstonite and paralstonite.
The spectrum is distinguishable from those of both polymorphs.
Compare spectrum with those of other members of aragonite group.

References:

1. Scheetz B.E. & White W.B. (1977)
   Vibrational spectra of the alkaline earth double carbonates.
   American Mineralogist, **62**, (1,2), pp.36-50.

2. Scheetz B.E. & White W.B. (1975)
   A vibrational study of the order/disorder in the alkaline earth double carbonates.
   Eos (Transactions of the American Geophysical Union. Washington), **56**(6), p.463.

Peak Table cm⁻¹

[3425]	867
2926	850
2587	839
2543	731
2501	722
2473	**700**
2362	**695**
2157	679
1793	**304**
1786	
1773	
1765	
1518	
1470	
1406	
1368	
1085	
1080	
878	

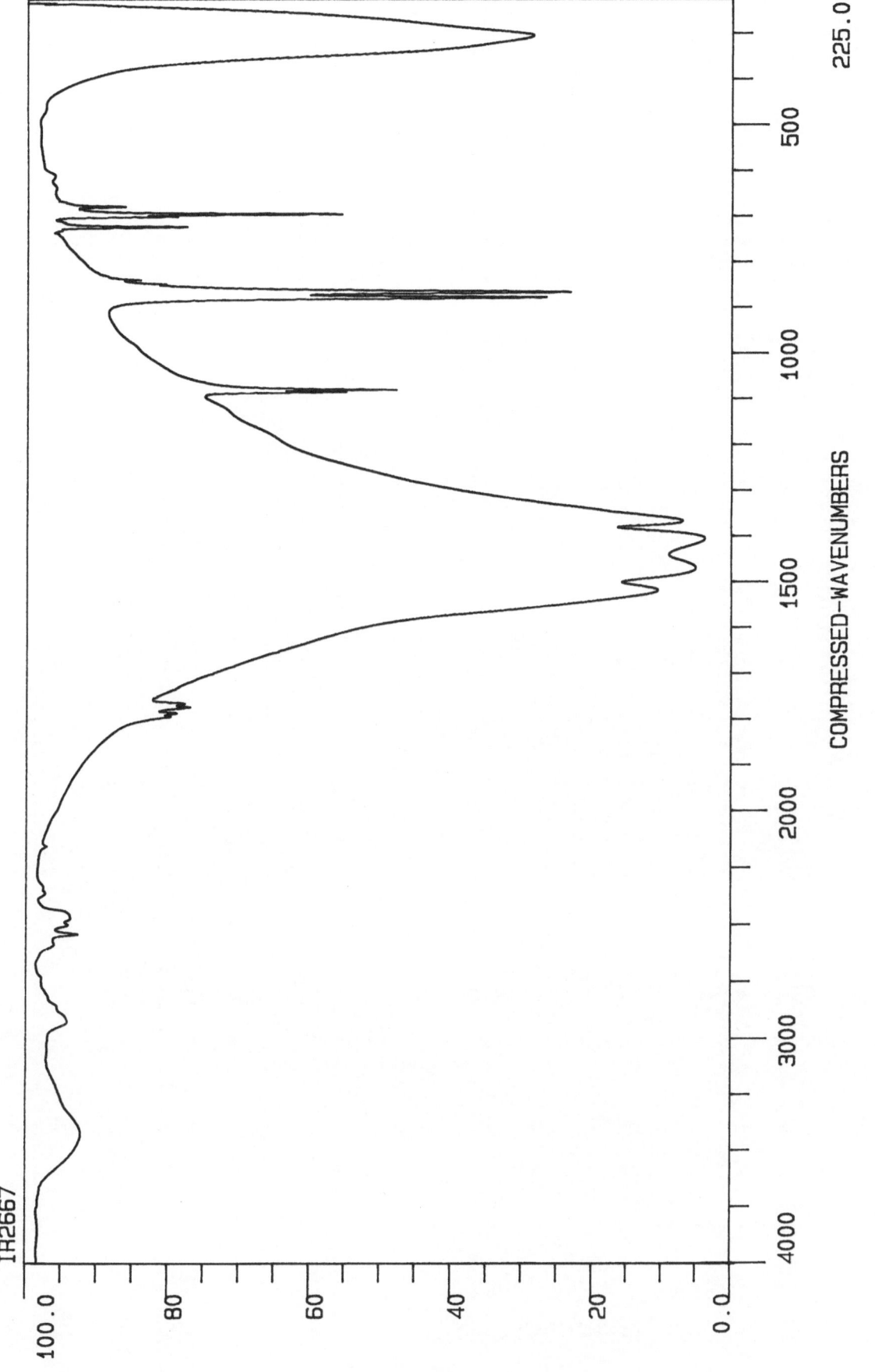

IR2667

% TRANSMITTANCE

COMPRESSED-WAVENUMBERS

BARYTOCALCITE

IR2667

% TRANSMITTANCE

100.0

80

60

40

20.0

1200

1000

WAVENUMBERS

800

600

BARYTOCALCITE [expanded detail]

BASTNÄSITE-(Ce)

Formula:	(Ce,La)(CO$_3$)F
Chemical class:	Anhydrous carbonate with hydroxyl or halogen
Chemical type :	(AB)(XO$_3$)Z$_q$

Crystal system:	Hexagonal
Mineral group:	Bastnäsite
Space group:	P$\bar{6}$2c

Specimen:	RMS unregistered
Source:	Tysfjord, Norway.
Spectrum ref. no.:	IR2958
Sample medium:	KBr disk
XRD:	4227
Composition:	Ce:La:Nd = 0·5:0·2:0·3 with minor Pr,Sm,Gd & Y

Notes

Forms a series with hydroxyl-bastnäsite-(Ce).
Spectra from a number of bastnäsite specimens from various localities were recorded. The spectra showed considerable variation, possibly due to (OH)/F substitution. IR2958 is an example of one of the least (OH)-bearing bastnäsite specimens studied, see **bastnäsite-(La)** for an (OH)-bearing example.

References:

1. Akhmanova M.N. & Orlova L.P. (1966)
   Investigation of rare-earth carbonates by infra-red spectroscopy.
   Geokhimiya, No.5, pp.71-578.
   Translated in: *Geochemistry International*, 3(3), pp.444-451.

2. Adler H.H. & Kerr P. F. (1963)
   Infrared spectra, symmetry and structure relations of some carbonate minerals.
   American Mineralogist, **48**, pp.839-853.

3. Donnay G. & Donnay J.D.H. (1953)
   The crystallography of bastnäsite, parisite, röntgenite and synchysite.
   American Mineralogist, **38**(11-12), pp.932-936.

Peak Table cm^{-1}

[3434]
2927
2858
2502
2342
1825
1760
1449
1089
868
842
763
736?
731
722?
612
366
275

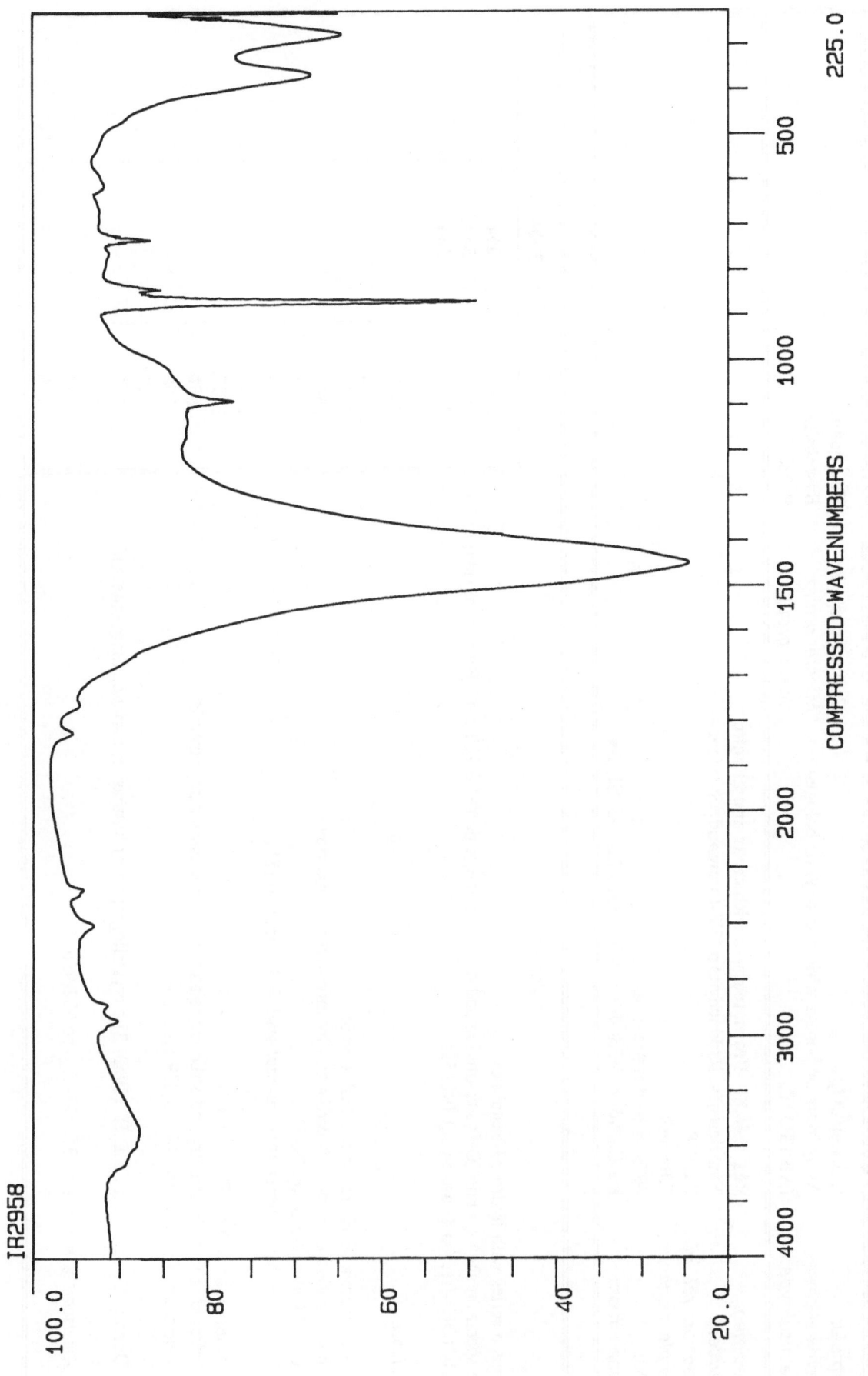

% TRANSMITTANCE

COMPRESSED-WAVENUMBERS

BASTNASITE- (Ce)

IR2958

BASTNÄSITE-(La)

Formula:	$(La,Ce)(CO_3)F$
Chemical class:	Anhydrous carbonate with hydroxyl or halogen
Chemical type:	$(AB)(XO_3)Z_q$
Crystal system:	Hexagonal
Mineral group:	Bastnäsite
Space group:	$P\bar{6}2c$

Specimen:	BM 1946,81 Brown massive with cerite and allanite
Source:	Nya Bastnäs, Riddarhyttan, Västmanland, Sweden.
Spectrum ref. no.:	IR2888
Sample medium:	KBr disk
XRD:	4976 = bastnäsite group
Composition:	La:Ce:Nd = 0·5:0·4:0·1 with trace Ca, Al, Si, Ba

Peak Table cm^{-1}

3747	**664**
3608	**360**
3581	**266**
3494	
3440	
2842	
2582	
2500	
1822	
1760	
1443	
1087	
880	
868	
842	
789	
749	
728	
720	

Notes

Forms a series with hydroxyl-bastnäsite.
The spectrum shows some (OH) substitution and differs in the region 900-700 cm^{-1} when compared to the nearly (OH)-free bastnäsite-(Ce) IR2958.

References:

1. Akhmanova M.N. & Orlova L.P. (1966)
   Investigation of rare-earth carbonates by infra-red spectroscopy.
   Geokhimiya, No.5, pp.71-578.
   Translated in: *Geochemistry International*, 3(3), pp.444-451.

2. Adler H.H. & Kerr P. F. (1963)
   Infrared spectra, symmetry and structure relations of some carbonate minerals.
   American Mineralogist, **48**, pp.839-853.

3. Donnay G. & Donnay J.D.H. (1953) The crystallography of bastnäsite, parisite, röntgenite and synchysite.
   American Mineralogist, **38**, (11-12), pp.932-936.

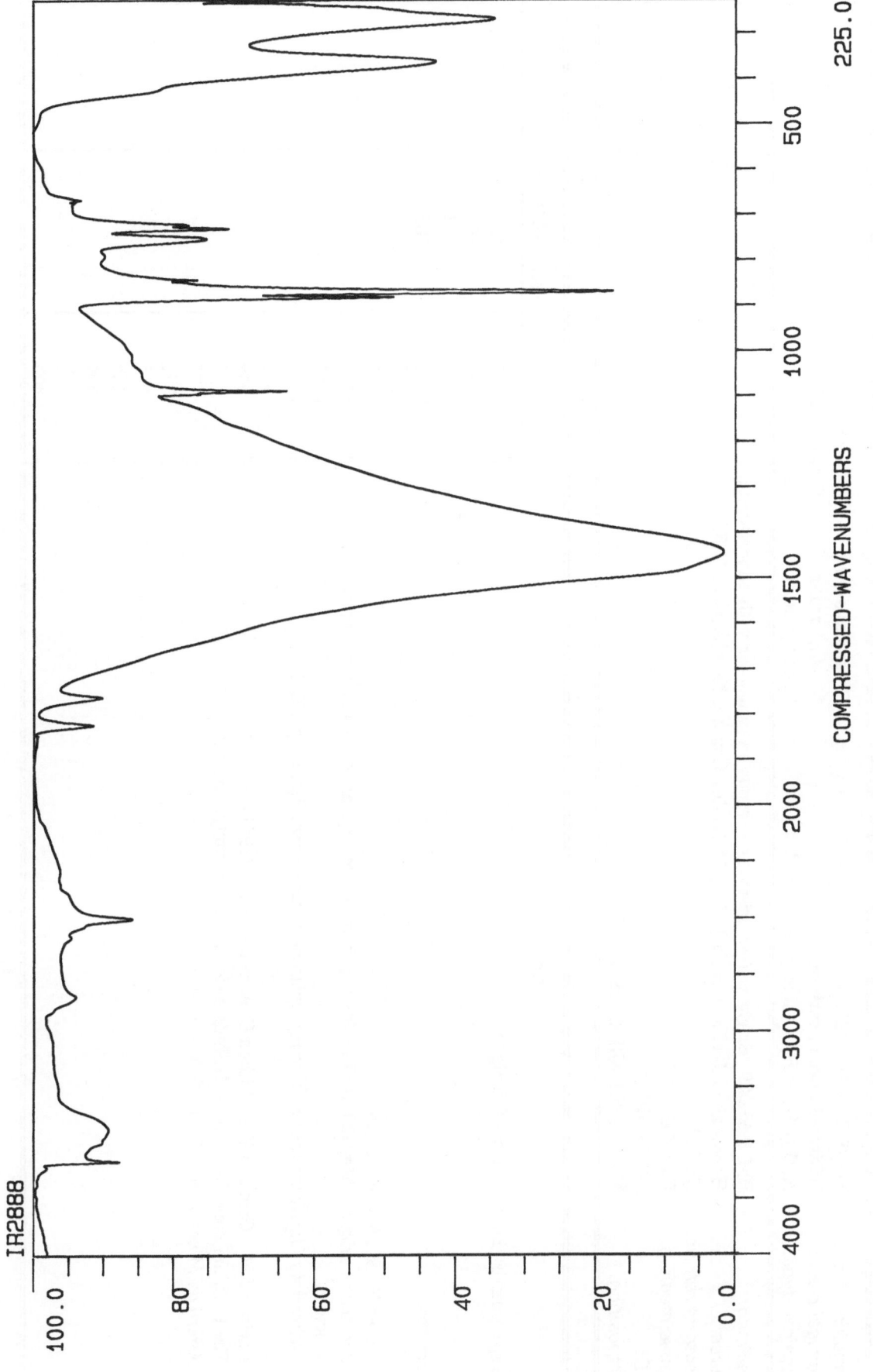

% TRANSMITTANCE

COMPRESSED–WAVENUMBERS

IR2888

225.0

BASTNASITE– (La)

BAYLEYITE

Formula:	$Mg_2(UO_2)(CO_3)_3 \cdot 18H_2O$
Chemical class:	Hydrated normal carbonate
Chemical type:	$A_mB_n(XO_3)_p \cdot xH_2O$ where (m+n):p = 1:1

Crystal system:	Monoclinic
Mineral group:	Rutherfordine
Space group:	$P2_1/a$

Specimen:	BM 1963,389 Bright yellow prismatic crystals on sandstone with anderstonite.
Source:	Homestake mine, Ambrosia Lakes, McKinley County, New Mexico, U.S.A.
Spectrum ref. no.:	IR2862
Sample medium:	KBr disk
XRD:	391F (std)
Composition:	Mg:U 1·5:1 with trace Si & S

Notes

The spectrum is similar to that of **liebigite**.

References:

1. Mayer H. & Mereiter K. (1986)
    Synthetic bayleyite, $Mg_2(UO_2)(CO_3)_3 \cdot 18H_2O$; thermochemistry, crystallography and crystal structure.
    Tschermaks Mineralogische und Petrographische Mitteilungen. **35**(2), pp.133-146.

2. Axelrod J.M., Grimaldi F.S., Milton C. & Murata K. J. (1951)
    The Uranium minerals from the Hillside mine, Yavapai County, Arizona.
    American Mineralogist, **36**, p.10.

Peak Table cm⁻¹

3547	426
3406	396
2233	374
2116	285
1619	
1553	
1387	
1144	
1116	
903	
849	
795	
775	
731	
693	
668	
604	
510	
465	

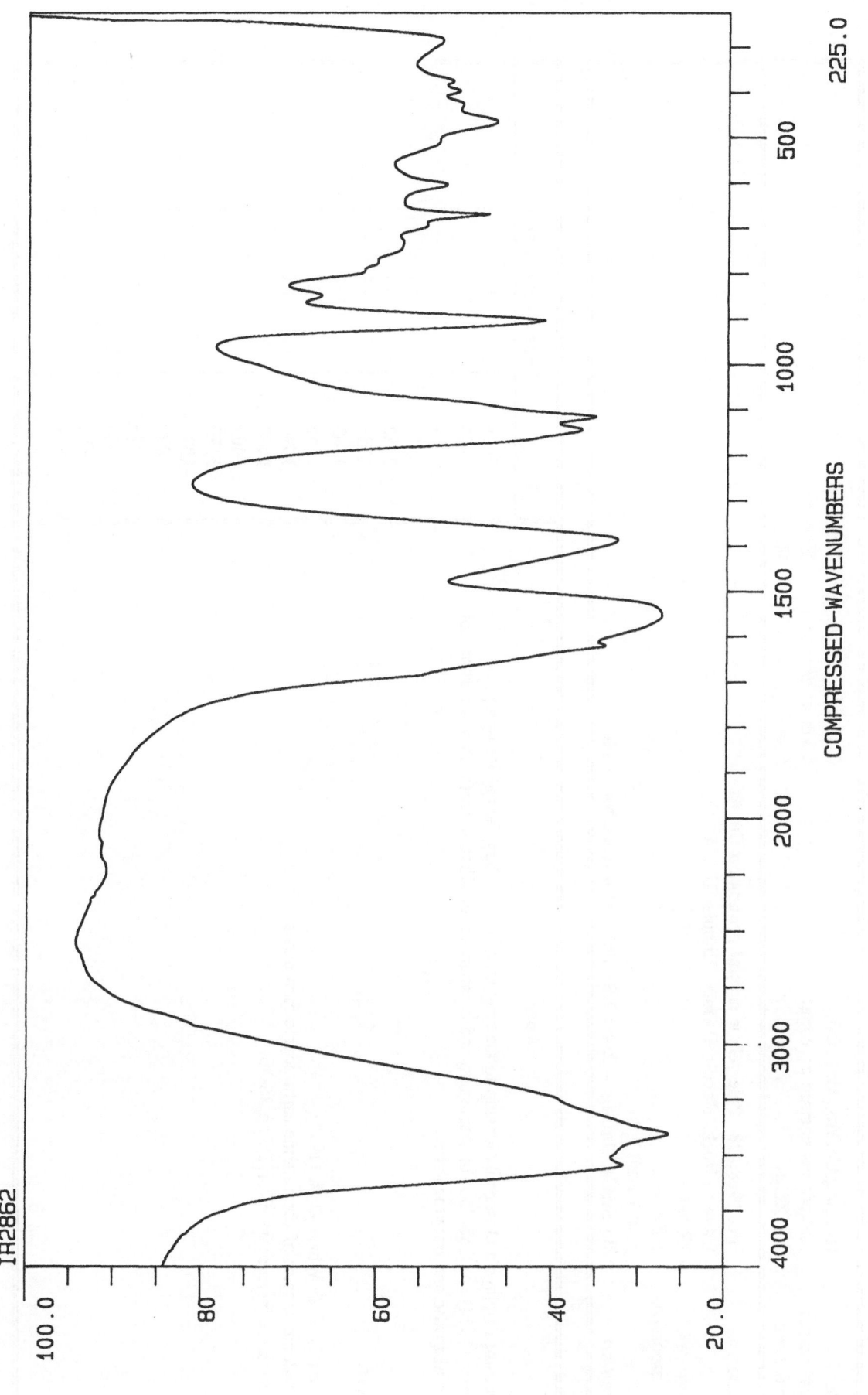

IR2862

% TRANSMITTANCE

COMPRESSED-WAVENUMBERS

BAYLEYITE

BENSTONITE

Crystal system:	Trigonal
Mineral group:	Huntite
Space group:	R$\bar{3}$

Formula:	(Ba,Sr)$_6$(Ca,Mn)$_6$Mg(CO$_3$)$_{13}$
Chemical class:	Anhydrous normal carbonate
Chemical type:	AB(XO$_3$)$_3$
Specimen:	BM 1968,628 Pale yellow crystal groups on fluorite.
Source:	Cave in Rock, Hardin County, Illinois, U.S.A.
Spectrum ref. no.:	IR2802
Sample medium:	KBr disk
XRD:	13487 (std)
Composition:	Ba:Sr:Ca:Mn:Mg = 2·4:0·2:3·1:0:1 with trace Na & Pb.

Peak Table cm^{-1}

[3428]	711
2923	700
2854	691
2510	684
2482	513
1781	465
1763	341
1755	295
1496	266
1447	
1409	
1179	
1087	
872	
845	
800	
780	
719	

Notes

The spectrum displays a closer relationship to the aragonite, rather than the calcite group.
See Farmer (1974) pp.258-259 for discussion and comparison of alkaline earth double carbonates.
Compare alstonite and barytocalcite.

References:

1. Scheetz B.E. & White W.B. (1977)
   Vibrational spectra of the alkaline earth double carbonates.
   American Mineralogist, 62(1,2), pp.36-50.

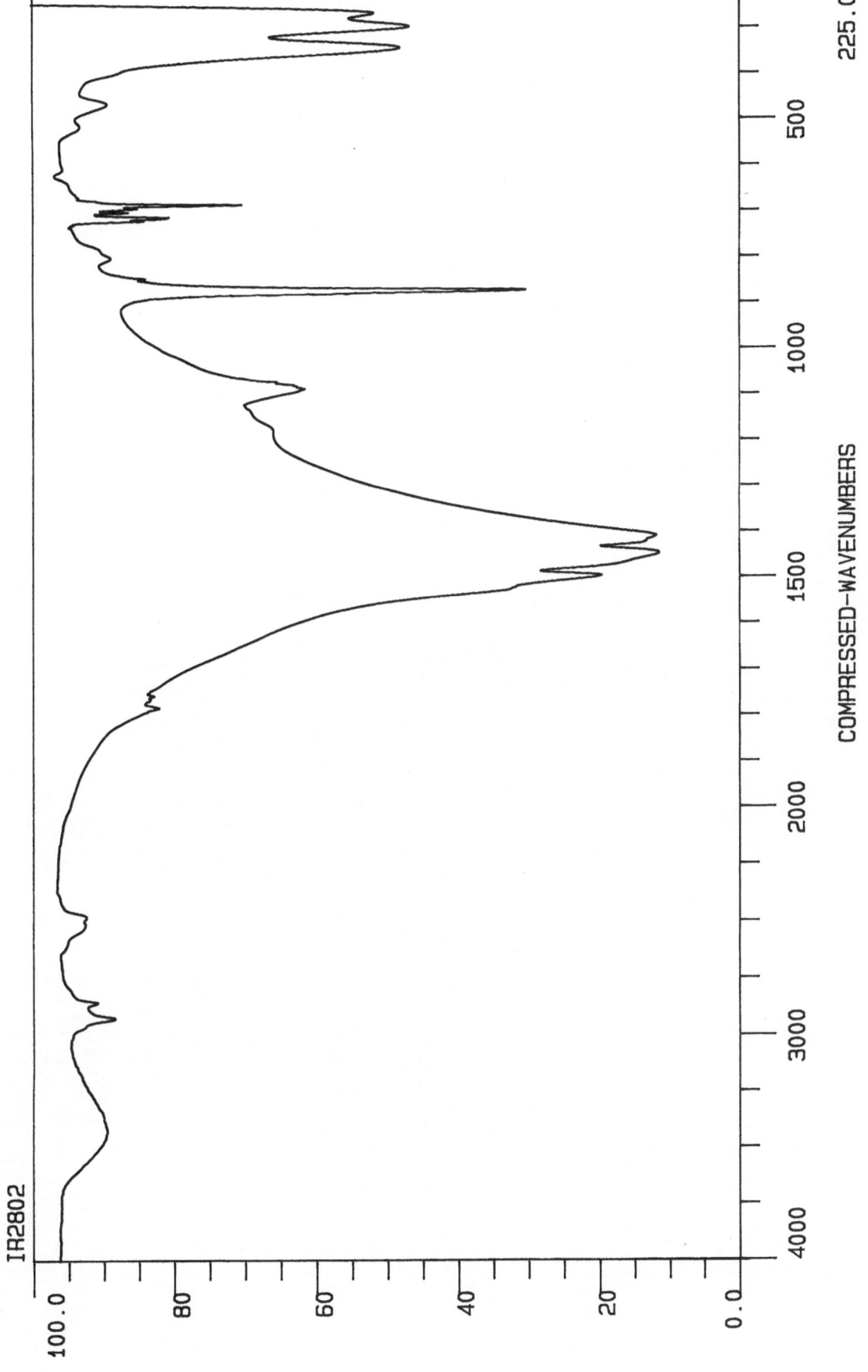

IR2802

% TRANSMITTANCE

100.0   80   60   40   20   0.0

4000   3000   2000   1500   1000   500   225.0

COMPRESSED-WAVENUMBERS

BENSTONITE

IR2802

% TRANSMITTANCE

WAVENUMBERS

BENSTONITE [expanded detail]

BEYERITE

Formula:	$(Ca,Pb)Bi_2(CO_3)_2O_2$ or $Ca(BiO_2)(CO_3)_2$	Crystal system:	Tetragonal
Chemical class:	Anhydrous carbonate with hydroxyl or halogen	Mineral group:	Bismutite
Chemical type:	$(AB)_3(XO_3)_2Z_q$	Space group:	I4/mmm

Specimen:	BM 1965,254 Grey/white compact with mica, chalcocite, malachite on garnet.
Source:	Meyer's Ranch pegmatite, Park Co., Colorado, U.S.A.
Spectrum ref. no.:	IR2883
Sample medium:	KBr disk
XRD:	11117 = beyerite
Composition:	

Peak Table cm^{-1}

	1032*
	1008*
	954*
	909*
3696?	861
3654?	848*
3621?	836?
3589?	756
3471?	707
[3324]	700
2926?	693
2852?	685
2381?	680
1764	632
1752	568
1746	471
1644	330
1564	
1482	
1431	
1197	
1100	
1065	

Notes

The spectrum is similar to that of **bismutite**.

* Peaks between 1032 and 909 cm^{-1} inc. are probably due to impurities.

Matches Suhner (5-65 A), beyerite, including the triplet at ≈ 1750 cm^{-1}.

References:

1. Kupcik V. (1979) Bismuth; crystal chemistry.
   In: Angino E.E & Long D.T. (Eds.) *Geochemistry of bismuth*. pp.13-19.
   Pub: Dowden, Hutchinson and Ross, Stroudsburg, PA, U.S.A.

2. Lagercrantz A. & Gunnar S.L. (1948)
   On the crystal structure of $Bi_2O_2CO_3$ (bismutite) and $CaBi_2O_2(CO_3)_2$ (beyerite).
   Arkiv för Kemi, Mineralogi och Geologi. (K. Svenska Vetenskapsakad), **25**(20).

3. Heinrich E.W. (1947) Beyerite from Colorado.
   American Mineralogist, **32**(11), pp.660-666.
   (contains analysis of material from Meyer's Ranch locality)

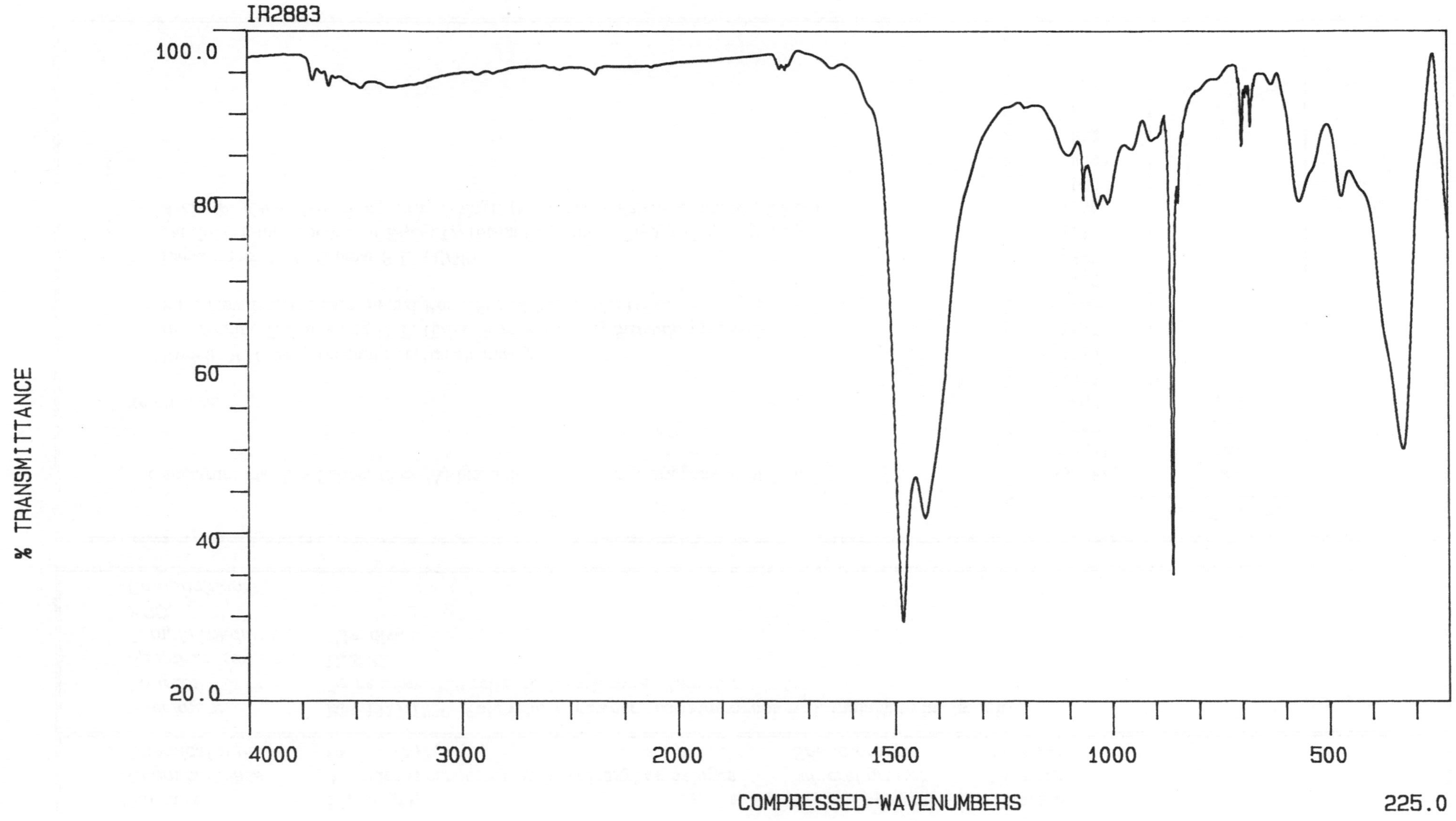

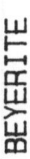

IR2883

100.0

80

60

40

20.0

% TRANSMITTANCE

4000　　　3000　　　2000　　　1500　　　1000　　　500　　　225.0

COMPRESSED-WAVENUMBERS

BEYERITE

BISMUTITE

Formula:	Bi$_2$(CO$_3$)O$_2$
Chemical class:	Anhydrous carbonate with hydroxyl or halogen
Chemical type:	(AB)$_2$(XO$_3$)Z$_q$

Crystal system:	Tetragonal
Mineral group:	Bismutite
Space group:	I4/mmm

Specimen:	BM 1929,1830 Pale yellow, powdery pseudomorphs with malachite in limonite.
Source:	Jessie mine, 120 miles S.E. of Kabwe, Zambia.
Spectrum ref. no.:	IR2882
Sample medium:	KBr disk
XRD:	
Composition:	

Peak Table cm^{-1}

[3468]	**668**	
2924?	**541**	
2852	**375**	
2404	**298**	
1755		
1734		
1645		
1560		
1455		
1393		
1132		
1066		
964		
889?		
862		
821		
759?		
691		

Notes

The spectrum matches Suhner (5-64 A) bismutite, but has an extra peak at 889 cm^{-1}.

References:

1. Kupcik V. (1979) Bismuth; crystal chemistry.
    In: Angino E.E & Long D.T. (Eds) *Geochemistry of bismuth*, pp.13-19.
    Pub: Dowden, Hutchinson and Ross, Stroudsburg, PA, USA.

2. Lagercrantz A. & Gunnar S.L. (1948)
    On the crystal structure of Bi$_2$O$_2$CO$_3$ (bismutite) and CaBi$_2$O$_2$(CO$_3$)$_2$ (beyerite).
    Arkiv för Kemi, Mineralogi och Geologi. (K. Svenska Vetenskapsakad), **25**(20).

IR2882

% TRANSMITTANCE

100.0   80   60   40   20   0.0

4000   3000   2000   1500   1000   500   225.0

COMPRESSED-WAVENUMBERS

BISMUTITE

BRENKITE

Formula:	**Ca$_2$(CO$_3$)F$_2$**
Chemical class:	**Anhydrous carbonate with hydroxyl or halogen**
Chemical type:	**(AB)$_2$(XO$_3$)Z$_q$**

Crystal system:	**Orthorhombic**
Mineral group:	
Space group:	**Pbcn**

Specimen:	**BM 1980,193** Tiny, radiating, colourless prismatic crystals on phillipsite.
Source :	**Schellkopf, nr. Brenk, Eifel, Germany** (type locality).
Spectrum ref. no.:	**IR2864**
Sample medium:	**KBr disk**
XRD:	**8109F (std)**
Composition:	**Ca & F only**

Peak Table cm^{-1}

[3316]	**718**
3006	**695**
2930	514
2856	460
2571	**355**
2505	**307**
2384 ?	
1808	
1524	
1506	
1455	
1189	
1165	
1087	
860	
843	
799	
780	
723	

Notes

The spectrum matches that given in the original description, ref. 3, except for the lack of a peak in the 600 cm^{-1} region.

References:

1. Leufer U. & Tillmanns E. (1980) Die Kristallstruktur von Brenkit, Ca$_2$F$_2$CO$_3$. *Tschermaks Mineralogische und Petrographische Mitteilungen*, **27**(4), pp.261-266. (in German with English summary).

2. Fleischer M., Chao G.Y. & Pabst A. (1979) New mineral names. *American Mineralogist*, **64**(12), pp.241-245.

3. Hentschel G., Leufer U. & Tillmanns E. (1978) Brenkit, ein neues Kalzium Fluor Karbonat vom Schellkopf Eifel. *Neues Jahrbuch für Mineralogie, Monatshefte*, **7**, pp.325-329. (German with English summary).

IR2864

% TRANSMITTANCE

100.0

80

60

40

20.0

4000

3000

2000

1500

1000

500

225.0

COMPRESSED-WAVENUMBERS

BRENKITE

BRUGNATELLITE

Formula:	$Mg_6Fe(CO_3)(OH)_{13}\cdot 4H_2O$
Chemical class:	Hydrated carbonate with hydroxyl or halogen
Chemical type:	Miscellaneous
Crystal system:	Trigonal
Mineral group:	Hydrotalcite (sjögrenite)
Space group:	P3 or P$\bar{3}$

Specimen:	BM 1910,560 Bronze, micaceous coating on serpentine.
Source:	Torre San Marino, Val Malenco, Valtellina, Lombardia, Italy.
Spectrum ref. no.:	IR2821
Sample medium:	KBr disk
XRD:	
Composition:	Mg:Fe = 4:1 + trace Mn, Ni

Peak Table cm^{-1}

3689	**1035**
3528	1028
3415	957
3292	**871**
3025	775
2359	722
2326	**674**
1680	**620**
1653	590
1548	**436**
1436	**382**
1418	310
1384	
1365	
1170	
1133	
1079	
1057	

Notes

The spectrum is similar to that of **sjögrenite** and other members of the **hydrotalcite** group.

References:

1. Bedogne F. & Pagano R. (1972)
    Mineral Collecting in Val Malenco.
    Mineralogical Record. 3(3), pp.120-123.

2. Fenoglio Massimo. (1938)
    Ricerche sulla brugnatellite.
    Periodico di Mineralogia. 9(1), pp.1-13.

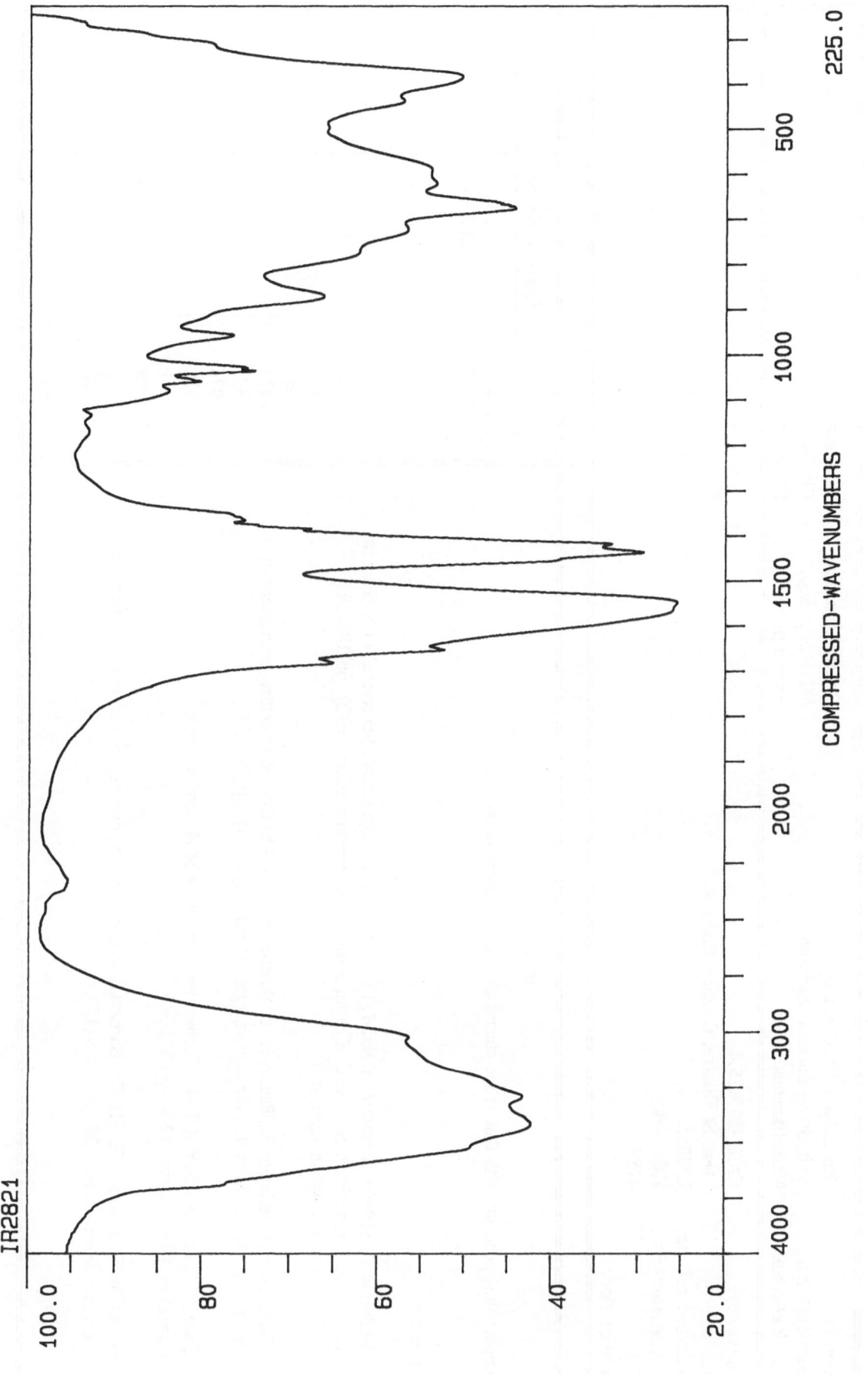

IR2821

% TRANSMITTANCE

COMPRESSED-WAVENUMBERS

BRUGNATELLITE

BURBANKITE

Formula:	(Na,Ca)$_3$(Sr,Ba,Ce)$_3$(CO$_3$)$_5$
Chemical class:	Anhydrous normal carbonate
Chemical type:	Miscellaneous

Crystal system:	**Hexagonal**
Mineral group:	**Eitelite**
Space group:	**P6$_3$/mmc**

Specimen:	RMS 1979.25.4
Source:	Mont St Hilaire, Quebec, Canada.
Spectrum ref. no.:	IR2912
Sample medium:	KBr disk
XRD:	4184
Composition:	

Notes

Compare the spectrum with that of the chemically similar **carbocernaite**.

References:

1. Ginderow D. (1989) Structure of Na$_3$M$_3$(CO$_3$)$_5$ (M = rare earth,Ca,Na,Sr) related to burbankite. *Acta Crystallographica, Section C, Crystal Structure Communications*, **45**(2), pp.185-187. (in French with English summary).

2. Effenberger H., Kluger F., Paulus H. & Woelfel E.R. (1985) Crystal structure refinement of burbankite. *Neues Jahrbuch für Mineralogie. Monatshefte*, (4), pp.161-170.

3. Chen T.T. & Chao G.Y. (1974) Burbankite from Mont St Hilaire, Quebec. *Canadian Mineralogist*, **12**(5), pp.342-345.

4. Pecora W.T. & Kerr J.H. (1953) Burbankite and calkinsite, two new minerals from Montana. *American Mineralogist*, **38**, pp.1169-1183.

Peak Table cm^{-1}

[3431]	
2925	
2858	
2487	
2362	
2334	
1773	
1498	
1451	
1411	466
1391	303
1076	
876	
861	
838	
742	
732	
712	
700	

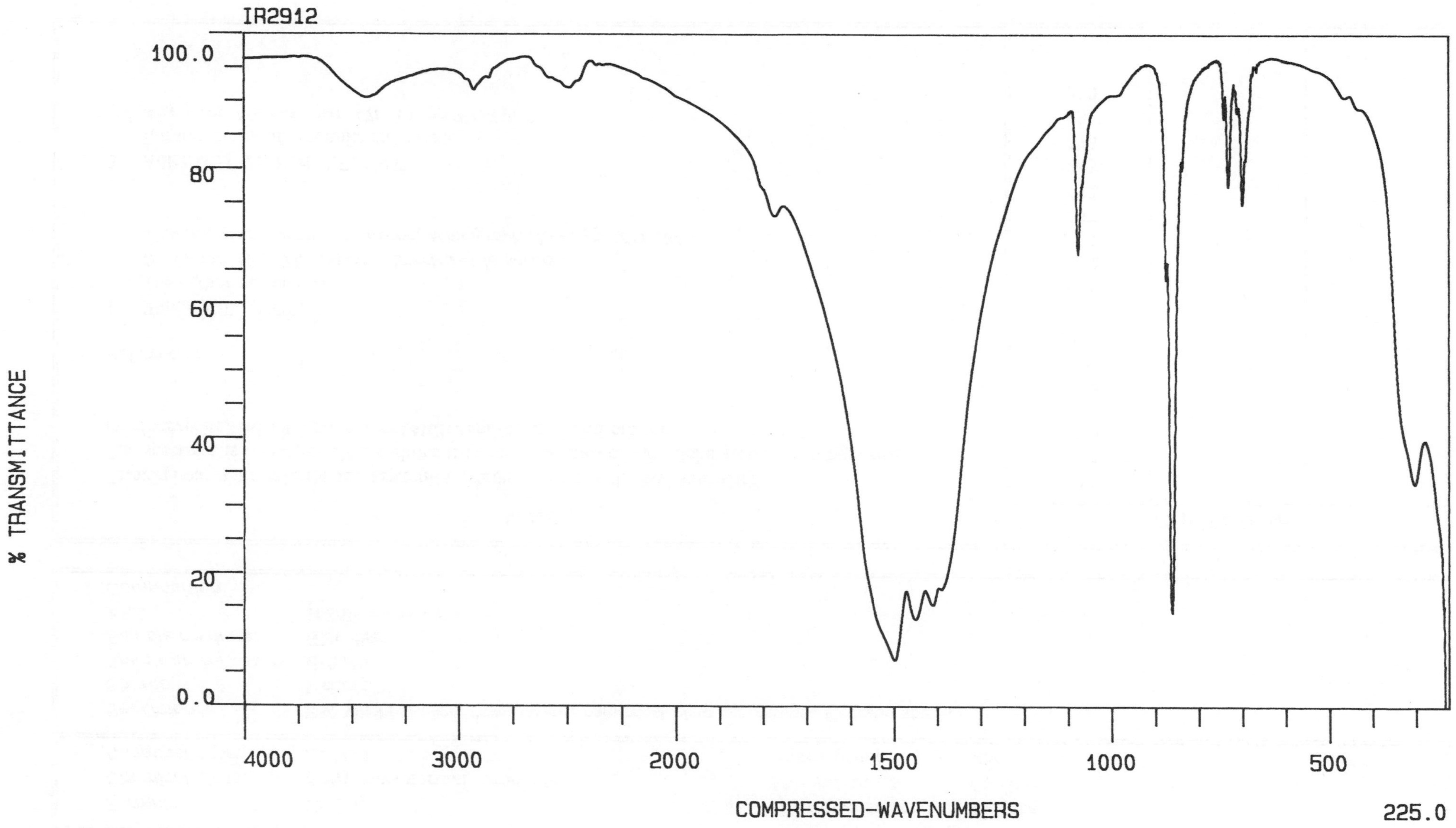

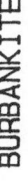

IR2912

% TRANSMITTANCE

100.0   80   60   40   20   0.0

4000   3000   2000   1500   1000   500   225.0

COMPRESSED–WAVENUMBERS

BURBANKITE

CALCITE

Formula:	CaCO$_3$	Crystal system:	Trigonal
Chemical class:	Anhydrous normal carbonate	Mineral group:	Calcite
Chemical type:	A(XO$_3$)	Space group:	R$\bar{3}$c

Specimen:	BM 90698 Large, transparent, colourless cleavage rhomb, (Iceland spar).
Source :	Iceland.
Spectrum ref. no.:	IR2600
Sample medium:	KBr disk
XRD:	2634F = calcite
Composition:	

Peak Table cm^{-1}

2980		
2931		
2872		
2587		
2515		
2170		
1798		
1734		
1429		
1162		
1012		
877		
848		
843?		
713		
482?		
320		

Notes

Trimorphous with **vaterite** and **aragonite**. Forms a series with **rhodochrosite**. The spectrum is typical of all anhydrous trigonal carbonates of the calcite group i.e. **magnesite, rhodochrosite, siderite, sphaerocobaltite, smithsonite and otavite.**

References:

1. White W.B. (1974)
    The carbonate minerals.
    In: Farmer (Ed) *The Infrared Spectra of Minerals.*
    Mineralogical Society of London, Monograph No.4, pp. 227-284.

2. Adler H.H. & Kerr P.F. (1962)
    Infrared study of aragonite and calcite.
    American Mineralogist, **47**(5,6), pp.700-717.

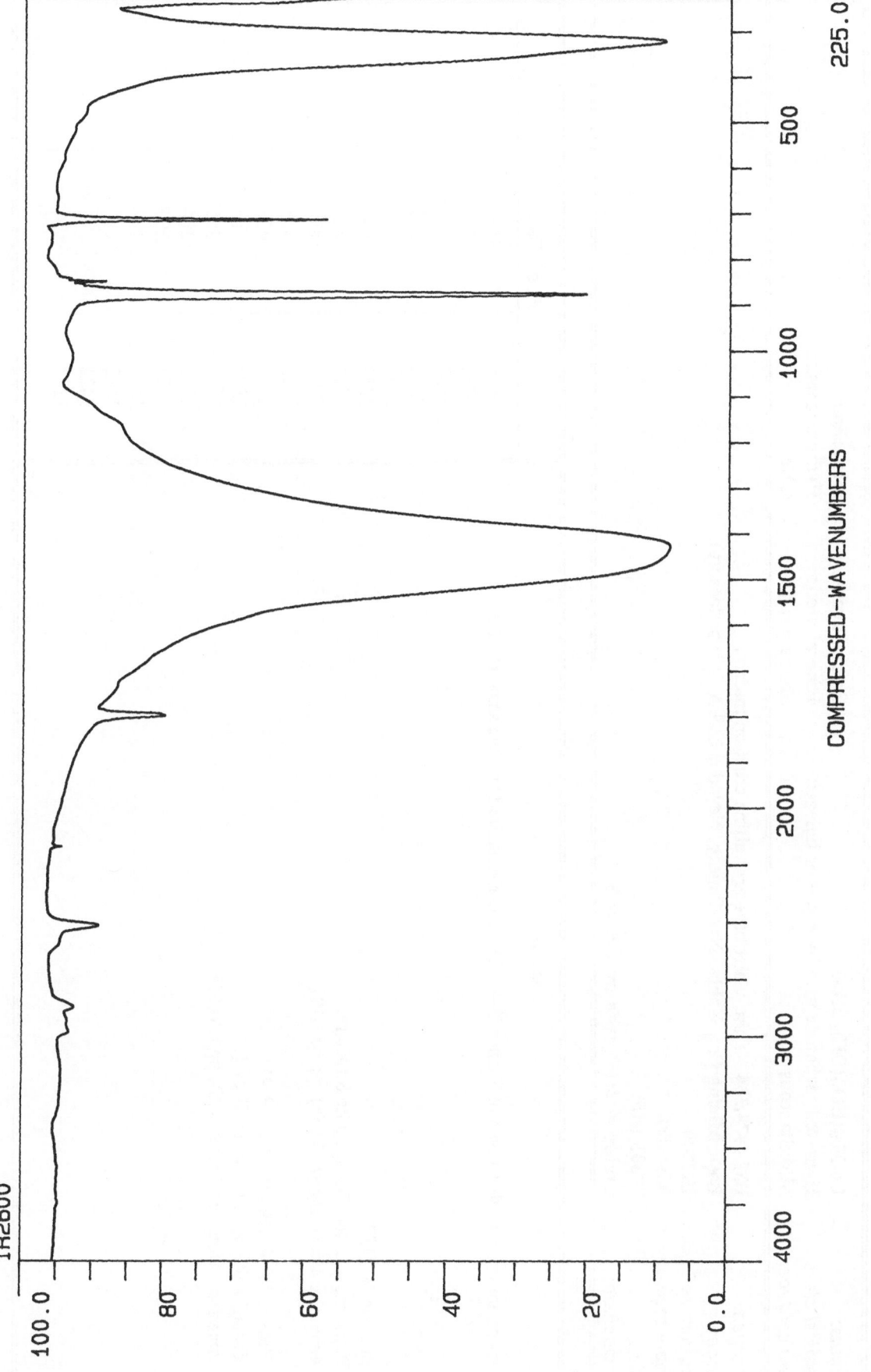

% TRANSMITTANCE

COMPRESSED-WAVENUMBERS

IR2600

CALCITE   var. Iceland Spar

CALLAGHANITE

Crystal system:	Monoclinic
Mineral group:	Hydromagnesite
Space group:	C2/c

Formula:	$Cu_2Mg_2(CO_3)(OH)_6 \cdot 2H_2O$
Chemical class:	Hydrated carbonate with hydroxyl or halogen
Chemical type:	Miscellaneous
Specimen:	BM 1978,334 Thin, violet/blue crystalline crust on matrix.
Source:	Basic Mining Co., Gabbs, Nye County, Nevada, U.S.A. (Type locality).
Spectrum ref. no.:	IR2800
Sample medium:	KBr disk
XRD:	7960F (std)
Composition:	Cu:Mg = 2:1·2 with trace Si & S

Peak Table cm^{-1}

3685	**1085**
3564	937
3507	855
3342	811
3244	**768**
3053	**686**
2921	**517**
2881?	**489**
2489	463
2366	**408**
1928	386
1845	355
1793	**304**
1627	250?
1567	
1462	
1422	
1171	

Notes

The spectrum matches that given in Suhner (5-41 A) for callaghanite from the same locality.

References:

1. Brunton G. (1973)
   Refinement of the Callaghanite Structure.
   American Mineralogist, **58**(56), pp.551-1973.

2. Beck C.W. & Burns J.H. (1954)
   Callaghanite, a new mineral [Nev.].
   American Mineralogist, **39**(7,8), pp.630-635.

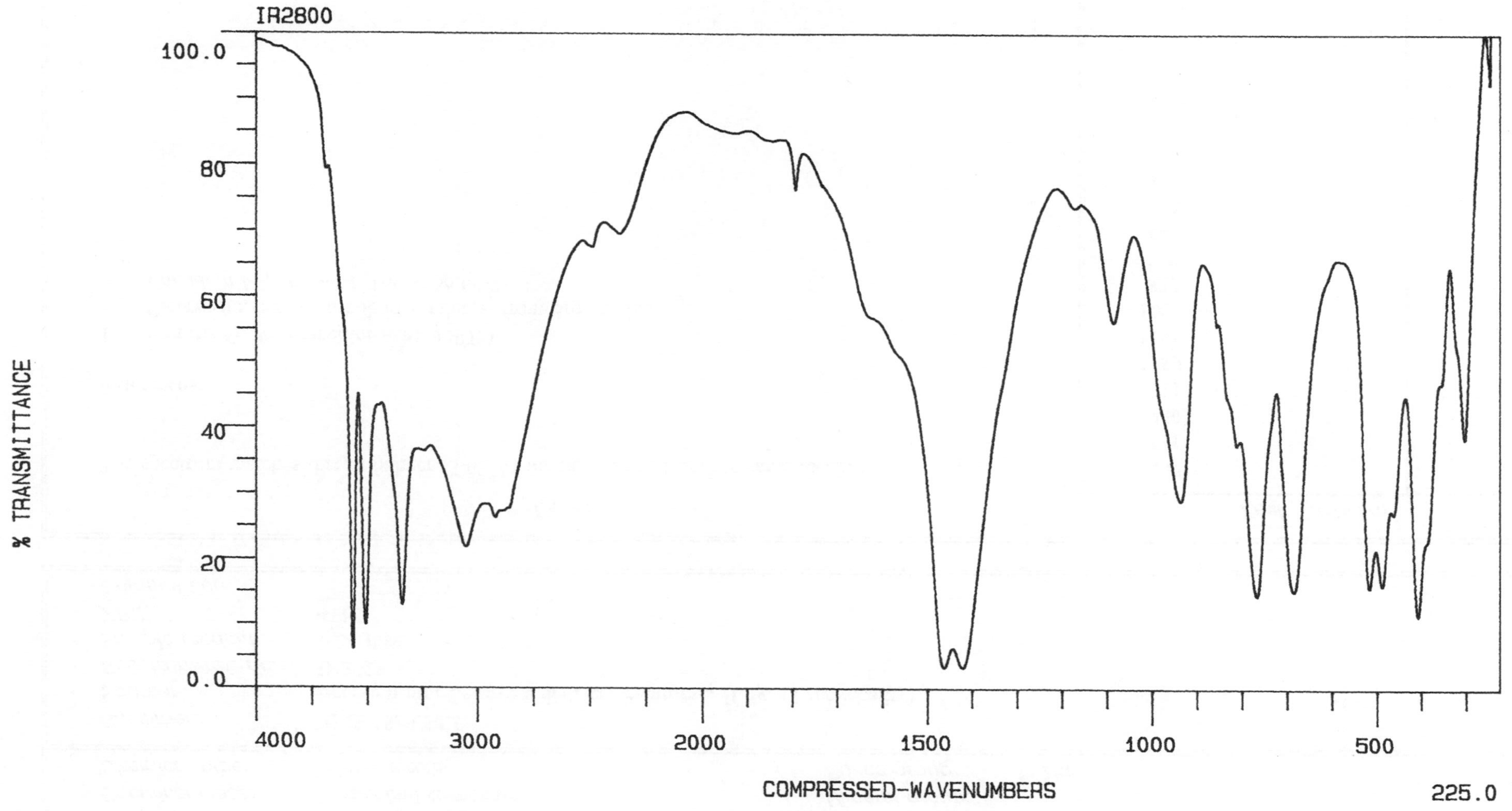

% TRANSMITTANCE

IR2800

100.0

80

60

40

20

0.0

4000

3000

2000

1500

1000

500

225.0

COMPRESSED-WAVENUMBERS

CALLAGHANITE

CANAVESITE

Formula:	Mg$_2$(CO$_3$)(HBO$_3$)·5H$_2$O
Chemical class:	Compound carbonate
Chemical type:	Miscellaneous
Crystal system:	Monoclinic
Mineral group:	
Space group:	2/m ?

Specimen:	RMS 1980.54.5.
Source:	Brosso mine, Canavese district, Piemonte, Italy. (Type locality).
Spectrum ref. no.:	IR2925
Sample medium:	KBr disk
XRD:	4118
Composition:	

Peak Table cm^{-1}

3407
1511
1440
1321
1159
1108
1006
877
768
693
440
339

Notes

The spectrum matches that in Suhner (5-63 A) for canavesite, from the same locality.

References:

1. Ferraris G. & Franchini A.M. (1978)
    Canavesite, a new carboborate mineral from Brosso, Italy.
    Canadian Mineralogist, 16(1), pp.69-73.

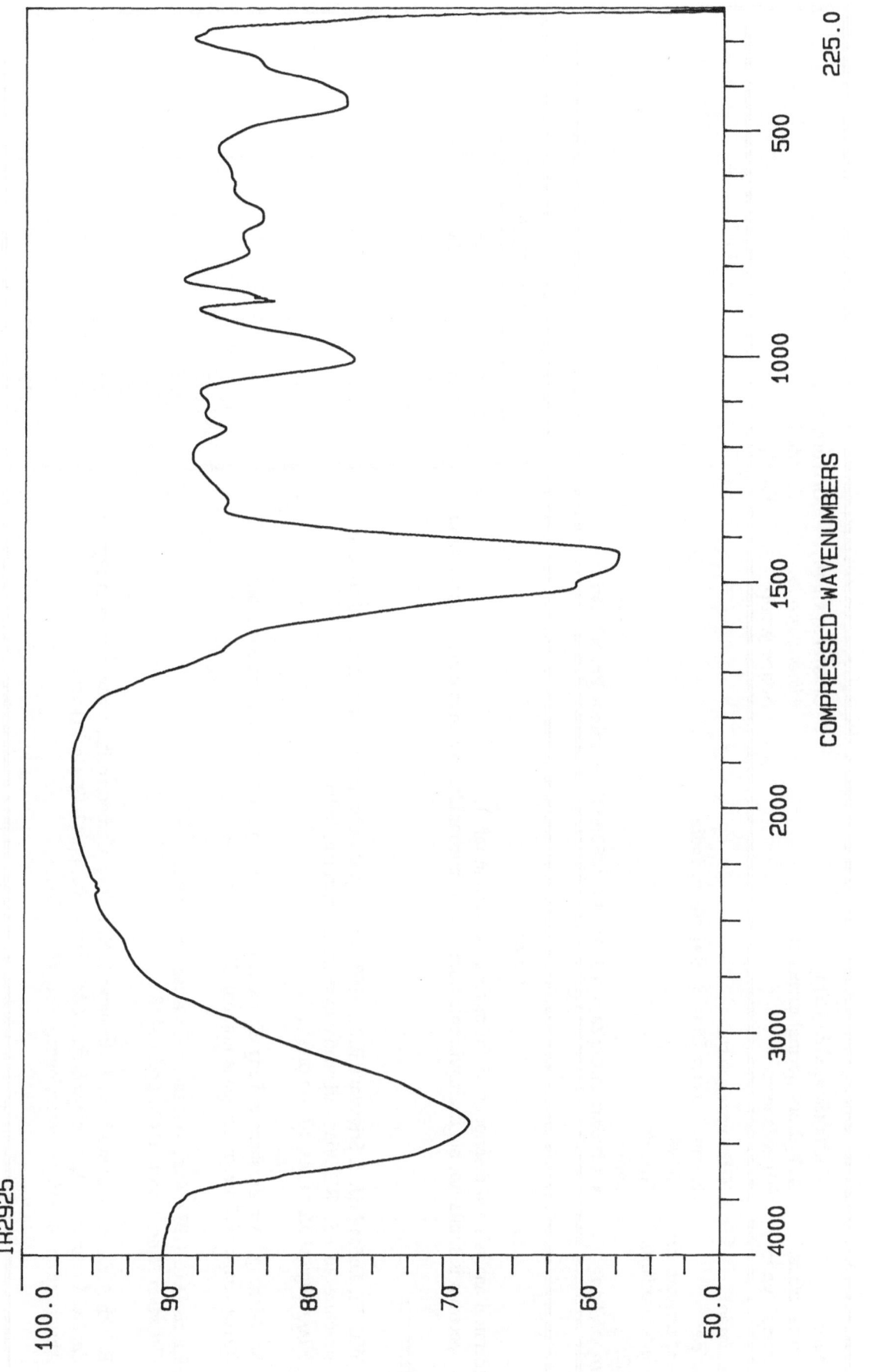

IR2925

% TRANSMITTANCE

100.0   90   80   70   60   50.0

4000   3000   2000   1500   1000   500   225.0

COMPRESSED-WAVENUMBERS

CANAVESITE

CARBOCERNAITE

Formula:	(Ca,Na)(Sr,Ce,Ba)(CO$_3$)$_2$
Chemical class:	Anhydrous normal carbonate
Chemical type:	Miscellaneous

Crystal system:	Orthorhombic
Mineral group:	Eitelite ?
Space group:	Pmc2$_1$

Specimen:	NHM unregistered.
Source:	Sarnu, Barmer District, Rajasthan, India.
Spectrum ref. no.:	IR2981
Sample medium:	KBr disk
XRD:	8661F
Composition:	Ca:Na:Sr:La:Ce:Ba = 1:0·4:0·4:0·2:0·2:0·03 + minor Pr, Nd, Sm

Notes

A chemical analysis and description of this material is given in ref. 1.
The spectrum is similar to, but distinguishable from that of **burbankite**, which is close in composition.

References:

1. Wall F., LeBas M.J. & Srivastava R.K. (1993) Calcite and carbocernaite exsolution and cotectic structures in a Sr, REE-rich carbonatite dyke from Rajasthan, India. *Mineralogical Magazine*, **57**, (in press).

2. Shi Nicheng, Ma Zhesheng & Peng Zhizhong (1982) The crystal structure of carbocernaite. *Kexue Tongbao* (Foreign Language Edition), **27**, (1), pp.76-80.

3. Harris D.C. (1972) Carbocernaite, a Canadian occurrence. *Canadian Mineralogist*, 11(4), pp.812-818.

4. Bulakh A.K., Kondrat'eva V.V. & Baranova E.N. (1961) Carbocernaite, a new rare earth carbonate. *Zapiski Vsesoiuznoe Mineralogicheskoe Obshchestvo*, **90**, pp.42-49. (In Russian) Abstracted in: *American Mineralogist*, 1961, **46**, p.1202.

Peak Table cm^{-1}

[3436]	573
2924	463
2860	322
2516	
1771	
1468	
1422	
1185	
1124	
1090	
1071	
873	
857	
802	
737	
716	
696	
637	
612	

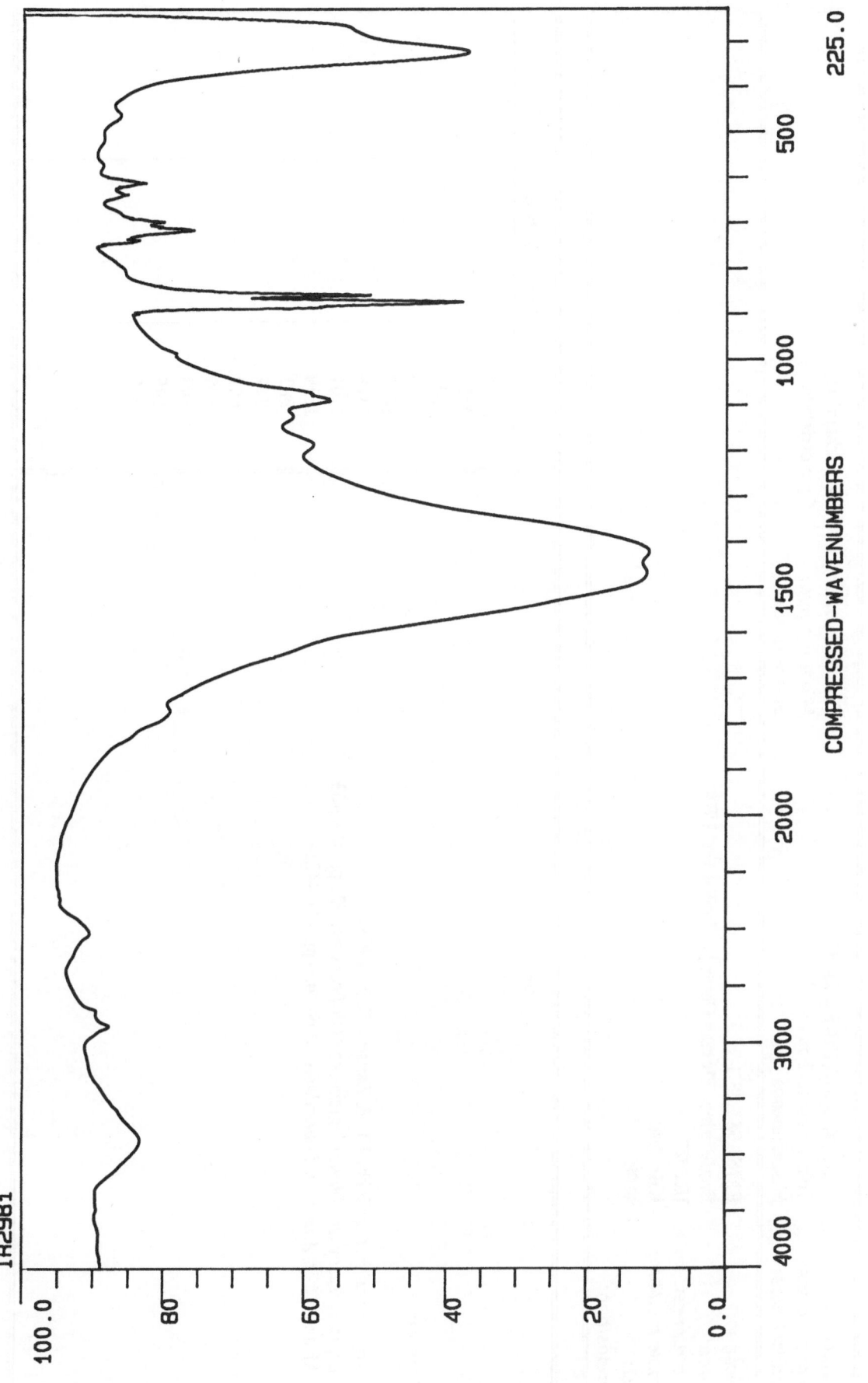

IR2981

% TRANSMITTANCE

100.0   80   60   40   20   0.0

COMPRESSED-WAVENUMBERS

4000   3000   2000   1500   1000   500   225.0

CARBOCERNAITE

CARBONATE-CYANOTRICHITE

Crystal system:	Orthorhombic
Mineral group:	Cyanotrichite
Space group:	?

Formula:	$Cu_4Al_2(CO_3,SO_4)(OH)_{12} \cdot 2H_2O$
Chemical class:	Compound carbonate
Chemical type:	Miscellaneous
Specimen:	RMS 1988.18.1.
Source:	Engle mine, Plumas County, California, USA.
Spectrum ref. no.:	IR2927
Sample medium:	KBr disk
XRD:	4108
Composition:	

Peak Table cm^{-1}

3416
2220
2056
1634
1453
1369
1101
1030
883
747
652
605
568
505
445

Notes

References:

1. Ankinovich E.A., Gekht I.I. & Zaitseva R.I. (1963)
   Zapiski Vsesoyuzni Mineralogicheskoe Obshchestva, **92**, pp.458–463.
   Abstracted in *American Mineralogist*, 1964, **49**, pp.441–442.

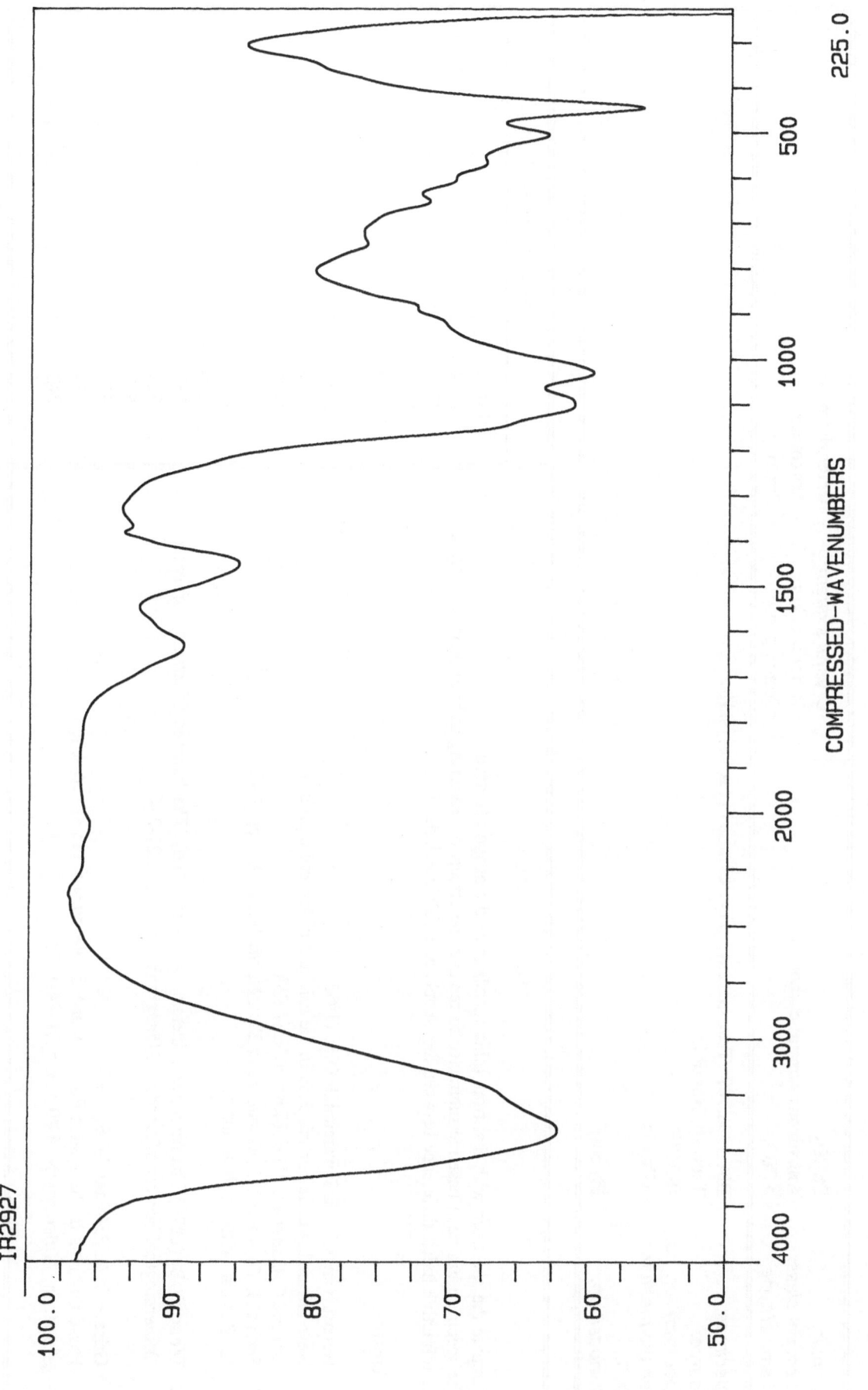

IR2927

% TRANSMITTANCE

100.0 90 80 70 60 50.0

4000 3000 2000 1500 1000 500 225.0

COMPRESSED-WAVENUMBERS

CARBONATE-CYANOTRICHITE

CERUSSITE

Formula:	PbCO₃
Chemical class:	Anhydrous normal carbonate
Chemical type:	A(XO₃)
Specimen:	BM 1926,187 Transparent, grey, twinned platy crystals.
Source:	Tsumeb, Namibia.
Spectrum ref. no.:	IR2679
Sample medium:	KBr disk
XRD:	
Composition:	Pb only

Crystal system:	Orthorhombic
Mineral group:	Aragonite
Space group:	Pmcn

Peak Table cm⁻¹

[3439]
2924
2730
2461
2404
1740
1727
1429
1395
1102
1051
994
839
824
698?
678
474
242?

Notes

Compare the spectrum with that from other members of the **aragonite** group.
The spectrum has been baseline subtracted to remove the effects of scattering at high wavenumbers due to the refractive index difference between the sample and KBr medium.

References:

1. Gevork'yan S.V. & Povarennikh O.S. (1983)
    New infrared spectra for minerals in the calcite and aragonite groups.
    Dopovidi Akademiyi Nauk Ukrayins'koyi RSR,
    Seriya B: Geologichni, Khimichni ta Biologichni Nauki. 11, pp.8-12.
    (In Russian with English summary).

2. White W.B. (1974) The carbonate minerals. *In:* Farmer (Ed), *The Infrared Spectra of Minerals.*
    Mineralogical Society of London, Monograph No. 4, pp.227-284.

3. Grisafe D.A. & White W.B. (1964)
    Phase relations in the system PbO CO₂ and the decomposition of cerussite.
    American. Mineralogist, **49**(9,10), pp.1184-1198.

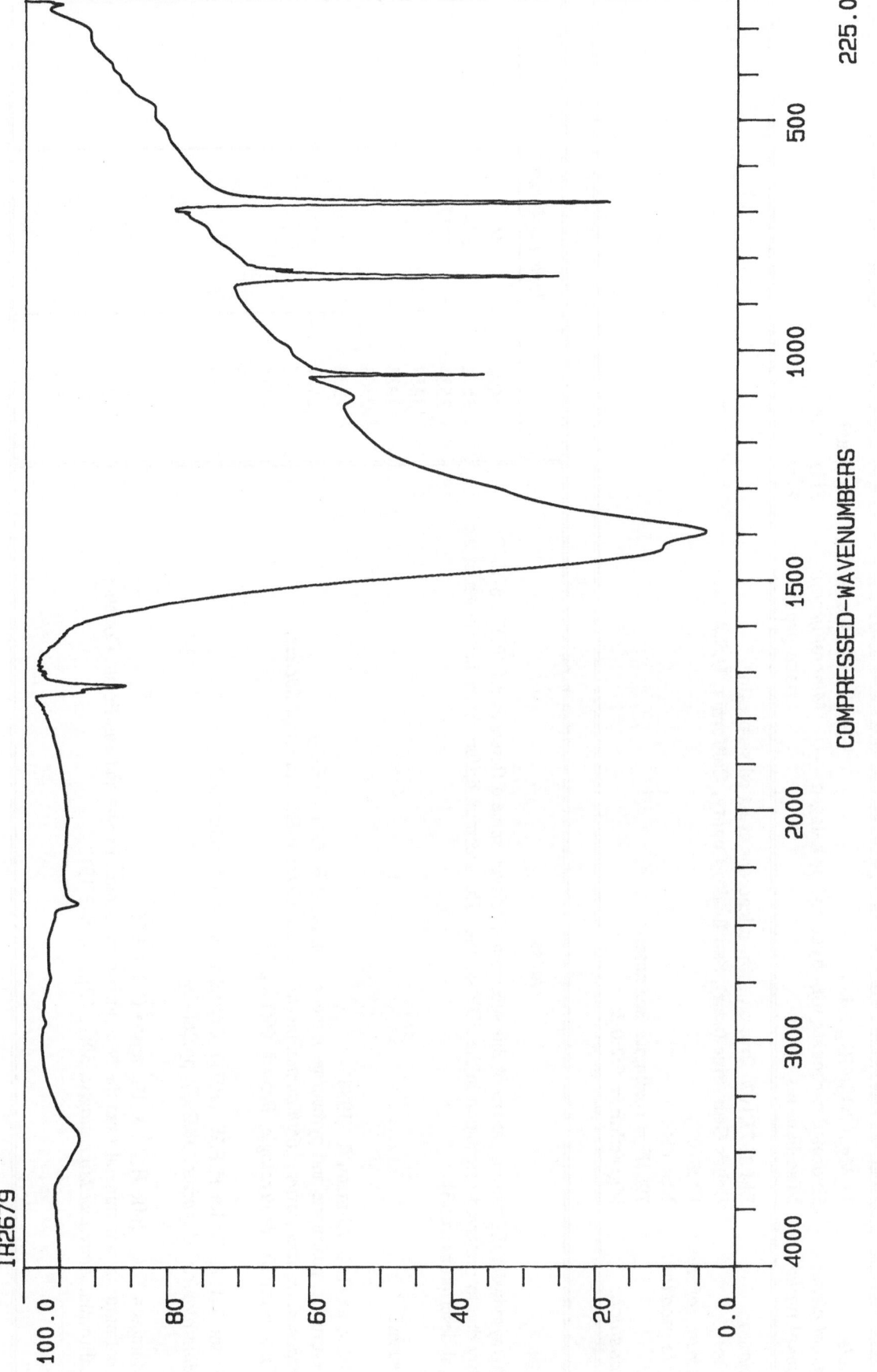

IR2679

% TRANSMITTANCE

COMPRESSED-WAVENUMBERS

CERUSSITE

COALINGITE

Formula:	$Fe_2Mg_{10}(CO_3)(OH)_{24}.2H_2O$
Chemical class:	Hydrated carbonate with hydroxyl or halogen
Chemical type:	**Miscellaneous**
Crystal system:	Trigonal
Mineral group:	Sjögrenite
Space group:	$R\bar{3}m$

Specimen:	BM 1977,102 Bronze, thin, micaceous crust on serpentine.
Source:	Dallas Gem mine (near), San Benito County, California, U.S.A.
Spectrum ref. no.:	IR2820
Sample medium:	KBr disk
XRD:	7962F = coalingite (see notes).
Composition:	Mg:Fe:Mn = 9:2:0·3

Peak Table cm⁻¹

3696		**441**
3641		376
3584		275?
3454		
2928		
2363		
2338		
1632		
1585		
1384		
1345		
1166		
1078		
1024		
958		
799		
778		
570		

Notes

The X-ray powder diffraction pattern of this specimen displayed some differences c.f. PDF 26-1217 possibly due to preferred orientation in the PDF sample. The spectrum differs from that shown in the original description (ref.3).

References:

1. Delnavaz H. & Allmann R. (1988)
   Fe-brucite, coalingite and pyroaurite in the system $MgO\text{-}Fe\text{-}O_2\text{-}H_2O(CO_2)$
   Proceedings and posters; 66th annual meeting of the German Mineralogical Society.
   Fortschritte der Mineralogie, Beiheft. **66**(1), p.23.

2. Pastor R.J. & Taylor H.F.W. (1971) Crystal structure of coalingite.
   Mineralogical Magazine, **38**(295), pp.286-294.

3. Mumpton F.A., Jaffe H.W. & Thompson C.S. (1965)
   Coalingite, a new mineral from the New Idria serpentinite, Fresno and San Benito Counties, California. *American Mineralogist,* **50**(11,12), pp.1893-1913.

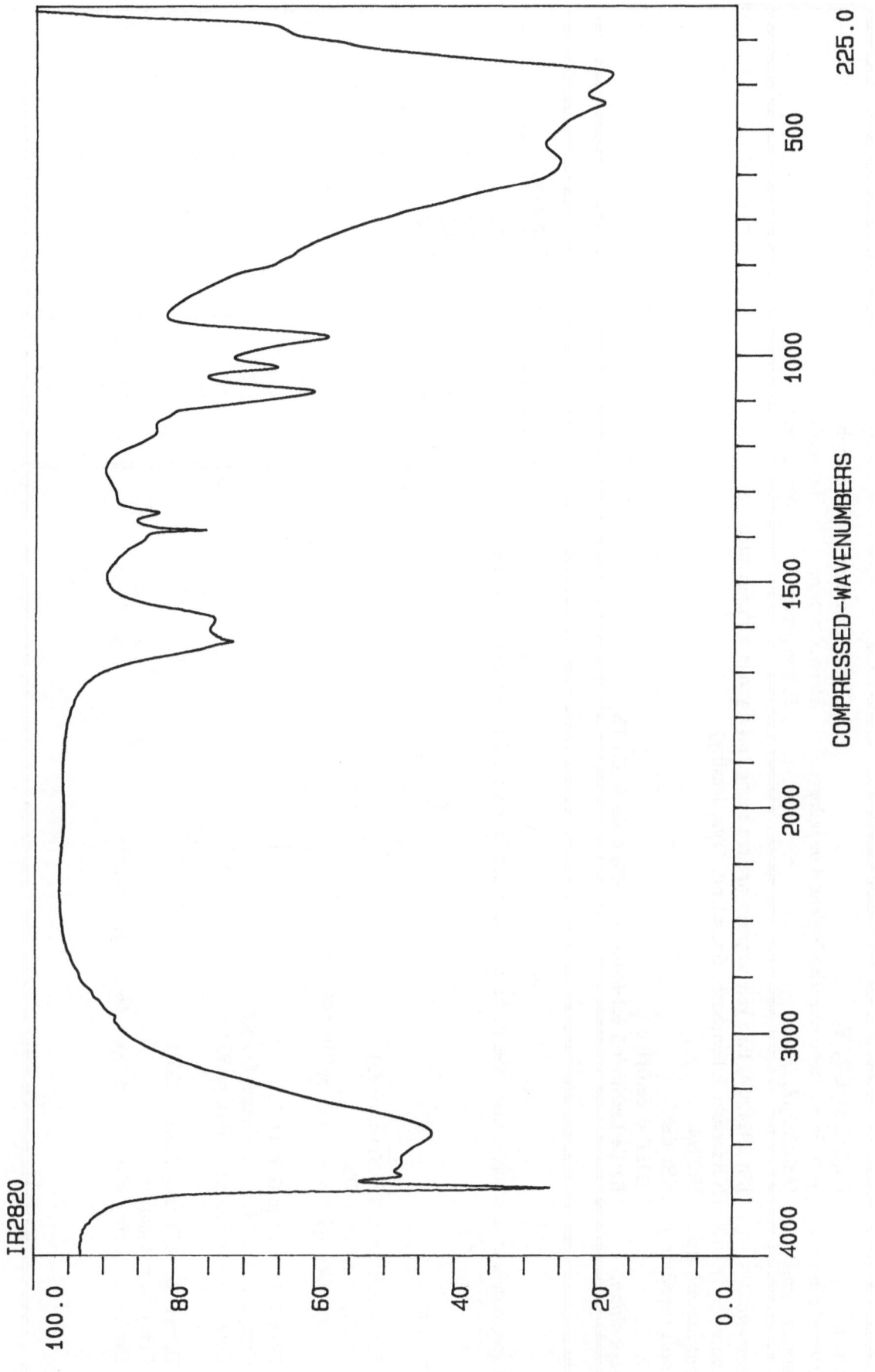

IR2820

% TRANSMITTANCE

COMPRESSED-WAVENUMBERS

COALINGITE

CORDYLITE-(Ce)

Formula:	$Ba(Ce,La)_2(CO_3)_3F_2$
Chemical class:	Anhydrous carbonate with hydroxyl or halogen
Chemical type:	$(AB)(XO_3)Z_q$

Crystal system:	Hexagonal
Mineral group:	Bastnäsite
Space group:	$P6_3/mmc$

Specimen:	BM 1924,854 Pale yellow/green striated hexagonal crystals with parisite.
Source:	Narsarsuk, Julianehaab, Greenland. (Type locality).
Spectrum ref. no.:	IR2894
Sample medium:	KBr disk
XRD:	8233F = cordylite
Composition:	Ba:Ce:La:Nd = $1 \cdot 0 : 1 \cdot 0 : 0 \cdot 6 : 0 \cdot 1$ with trace Sr,Ca,Th

Peak Table cm^{-1}

[3441]	513
2811	464
2612	405
2528	250
2476	
1808	
1802	
1774	
1486	
1408	
1180	
1091	
881	
857	
802	
719	
690	
634	

Notes

The spectrum is more complex than those of the chemically similar bastnäsite, synchysite and parisite.

References:

1. Zhang Peishan & Tao Kejie. (1985)
   Cordylite in Bayun Obo.
   Scientia Geologica Sinica, (2), pp.191-195.

2. Chen T.T. & Chao G.Y. (1975)
   Cordylite from Mont St. Hilaire, Quebec.
   Canadian Mineralogist, 13(1), pp.93-94.

3. Donnay G. & Donnay J.D.H. (1955)
   Cordylite re-examined.
   Geological Society of America Bulletin, 66(12), pt 2, p.1551.

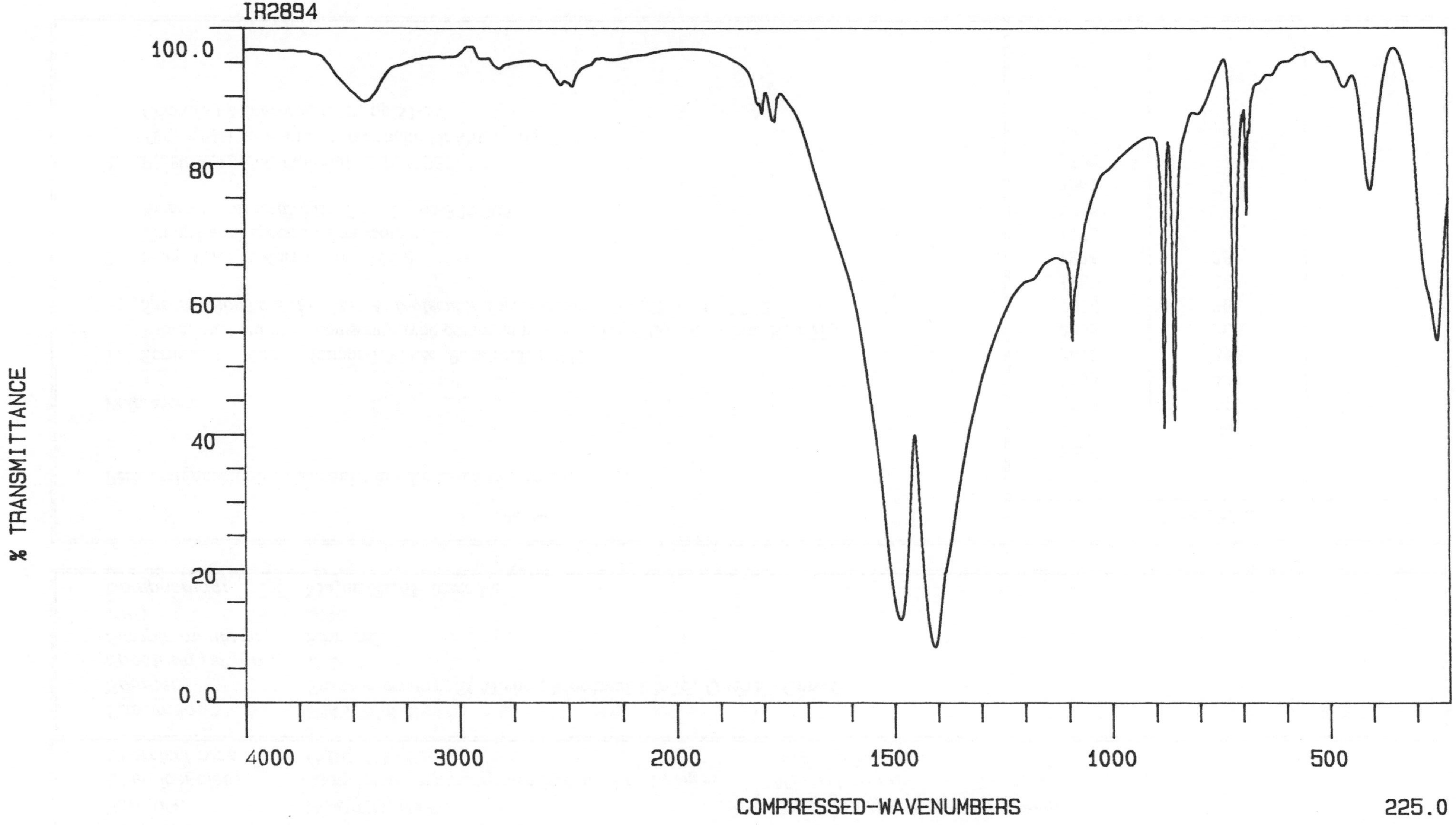

IR2894

% TRANSMITTANCE

100.0   80   60   40   20   0.0

4000   3000   2000   1500   1000   500   225.0

COMPRESSED-WAVENUMBERS

CORDYLITE- (Ce)

DAWSONITE

Formula:	NaAl(CO$_3$)(OH)$_2$
Chemical class:	Anhydrous carbonate with hydroxyl or halogen
Chemical type:	(AB)$_2$(XO$_3$)Z$_q$
Crystal system:	Orthorhombic
Mineral group:	Dawsonite
Space group:	Imam

Specimen:	RMS 1978.2.
Source:	Francon quarry, St Michel, Montreal Island, Quebec, Canada.
Spectrum ref. no.:	IR2935
Sample medium:	KBr disk
XRD:	4096
Composition:	Major Na,Al trace Fe

Peak Table cm^{-1}

3471	731
3286	694
2819	549
2758	517
2605	487
2472	394
2098	363
1978	306
1889	284
1826	252
1773	
1720	
1561	
1398	
1097	
953	
863	
848	

Notes

Peak assignments for dawsonite are discussed in Farmer.

References:

1. Serna C.J., Garcia Ramos J.V. & Pena M.J. (1985)
   Vibrational study of dawsonite type compounds MAl(OH)$_2$ CO$_3$ (M = Na, K, NH$_4$)
   Spectrochimica Acta, Part A: Molecular Spectroscopy, **41**, (5), pp.697-702.

2. Estep P.A. & Karr C. Jr. (1968)
   The infrared spectra of dawsonite.
   American Mineralogist, **53**(1,2), pp.305-309.

3. Frueh A.J.Jr. & Golightly J.P. (1967)
   The crystal structure of dawsonite NaAl(CO$_3$)(OH)$_2$
   Canadian Mineralogist, **9**, pp.51-56.

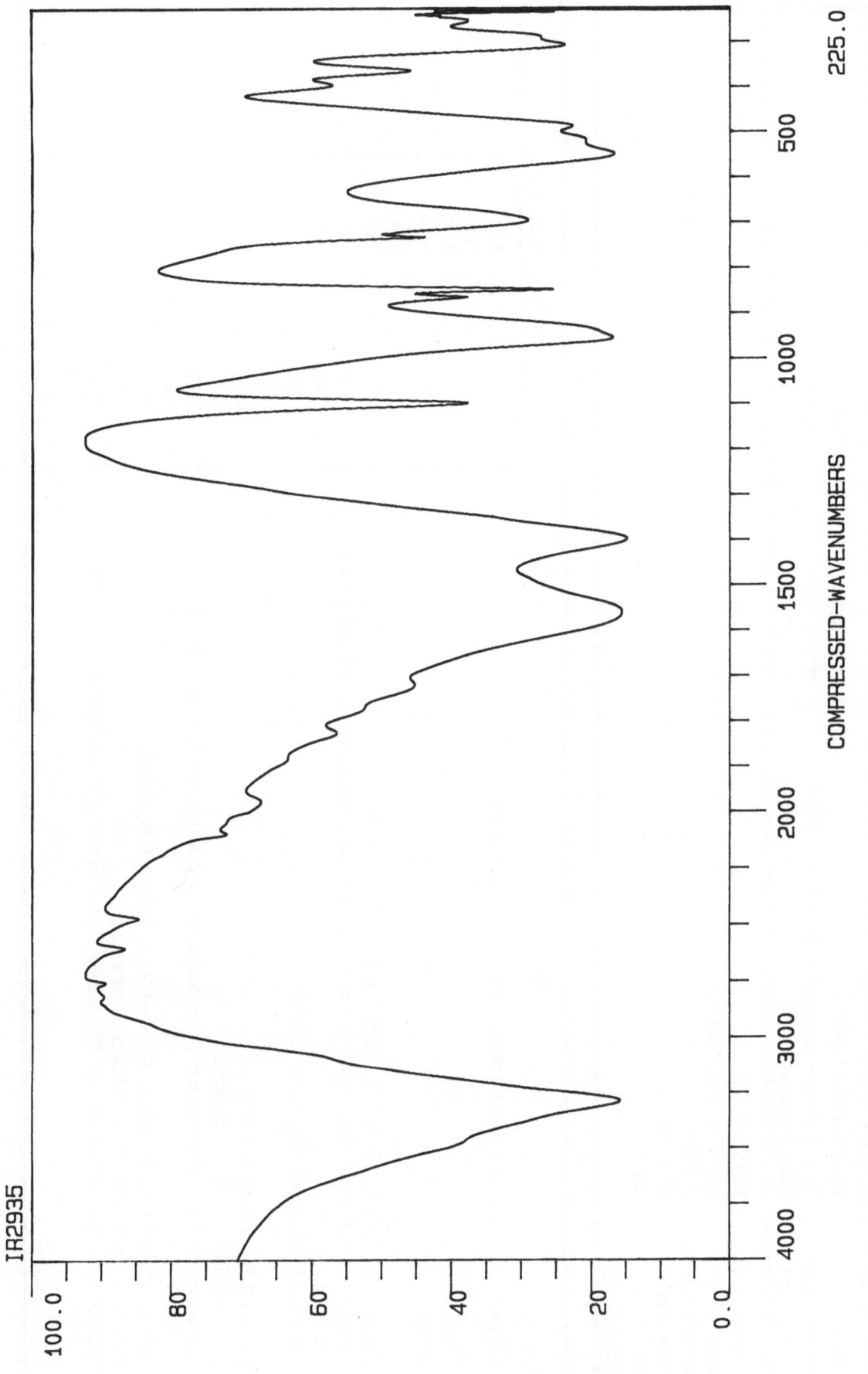

IR2935

% TRANSMITTANCE

COMPRESSED-WAVENUMBERS

DAWSONITE

DEFERNITE

Formula:	**Ca$_3$CO$_3$(OH,Cl)$_4$·H$_2$O**
Chemical class:	**Hydrated carbonate with hydroxyl or halogen**
Chemical type:	**Miscellaneous**

Crystal system:	**Orthorhombic**
Mineral group:	**Hydromagnesite**
Space group:	**Pna2$_1$ or Pnam**

Specimen:	**RMS 1988.7.7.**
Source:	**Kombat mine, Namibia.**
Spectrum ref. no.:	**IR2913**
Sample medium:	**KBr disk**
XRD:	**4101**
Composition:	

Peak Table cm^{-1}

3575	**859**
2927	**766?**
2564	**748?**
2481	**738?**
2363	**654**
2334	**538**
1777	**510**
1542	**343**
1467	
1414	
1303	
1255	
1211	
1187	
1080	
1039	
990	
934	
871	

Notes

See ref. no.1 for a discussion of the mineral formula and substitutions.

References:

1. Peacor D.R., Sarp H., Dunn P.J., Innes J. & Nelen J.A. (1988) Defernite from Kombat Mine, Namibia; a second occurrence, structure refinement, and crystal chemistry. *American Mineralogist,* **73**(78), pp.888-893.

2. Liebich B.W. & Sarp H. (1985) The crystalline structure of defernite. *Schweizerische Mineralogische und Petrographische Mitteilungen,* **65**(2,3), pp.153-158.

3. Sarp H., Taner M.F., Deferne J., Bizouard H. & Liebich B.W. (1980) Defernite, a new chloro-hydroxyl calcium carbonate. *Bulletin de la Société Francaise de Minéralogie et de Christallographié,* **103**(2), pp.185-189.

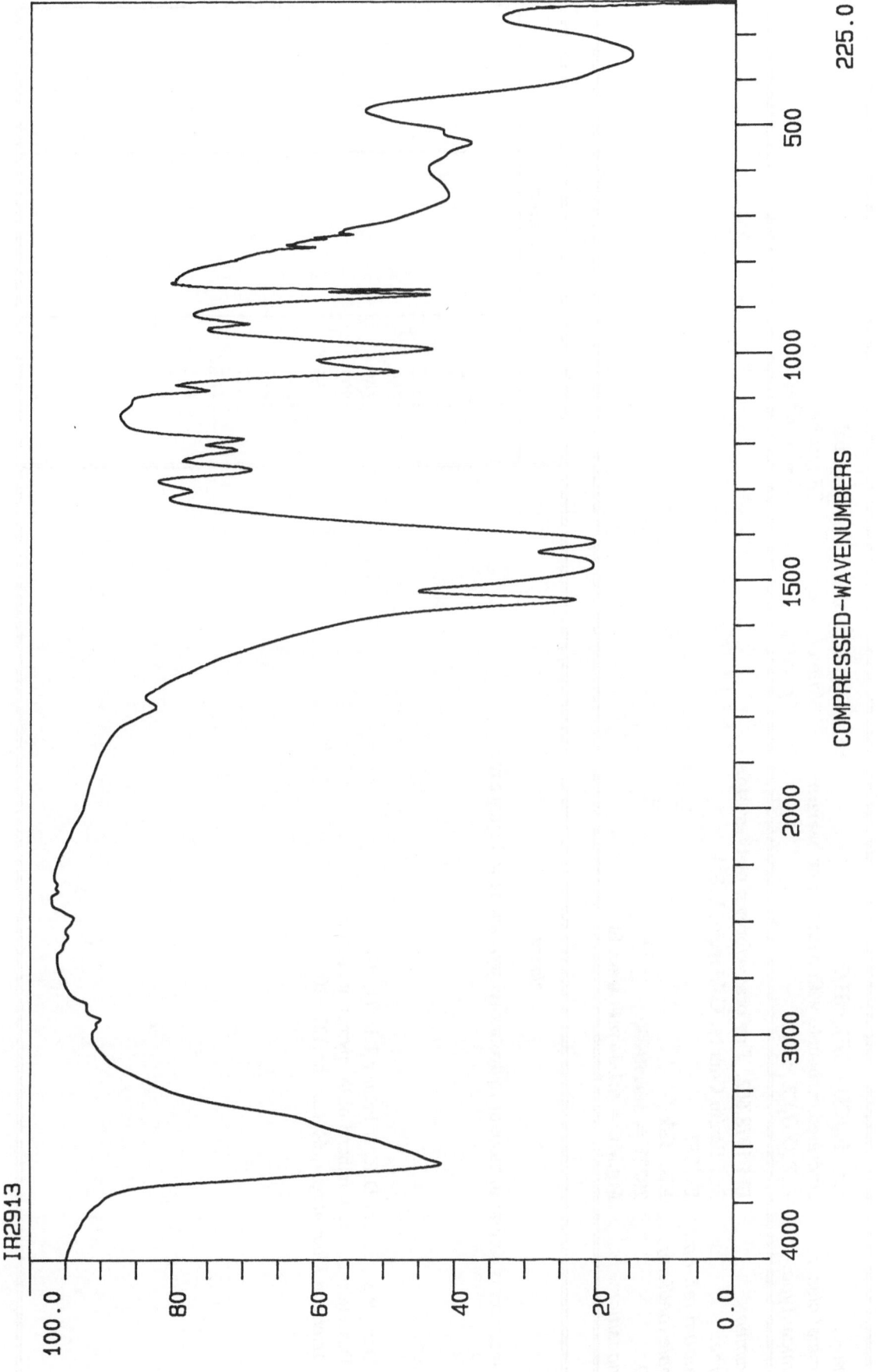

IR2913

% TRANSMITTANCE

COMPRESSED-WAVENUMBERS

DEFERNITE

DESAUTELSITE

Formula:	$Mg_6Mn_2(CO_3)(OH)_{16} \cdot 4H_2O$
Chemical class:	Hydrated carbonate with hydroxyl or halogen
Chemical type:	$A_mB_n(XO_3)_pZ_q \cdot xH_2O$
Crystal system:	Trigonal
Mineral group:	Sjögrenite
Space group:	$R\bar{3}m$ or R3m

Specimen:	BM 1978,602. Tiny orange/brown platy crystals.
Source:	San Benito County, California, U.S.A.
Spectrum ref. no.:	IR2857
Sample medium:	KBr disk
XRD:	20221 = desautelsite
Composition:	Mg:Mn = 6:1·3 with trace Si.

Peak Table cm⁻¹

3591
3452
1632
1595
1380
1349
1292
1156
1080
1017
997
661
613
401

Notes

The spectrum is similar to those of hydrotalcite, pyroaurite and sjögrenite.

References:

1. Dunn P.J., Peacor D.R. & Palmer T.D. (1979)
   Desautelsite, a new mineral of the pyroaurite group.
   American Mineralogist, 64(1,2), pp.127-130.

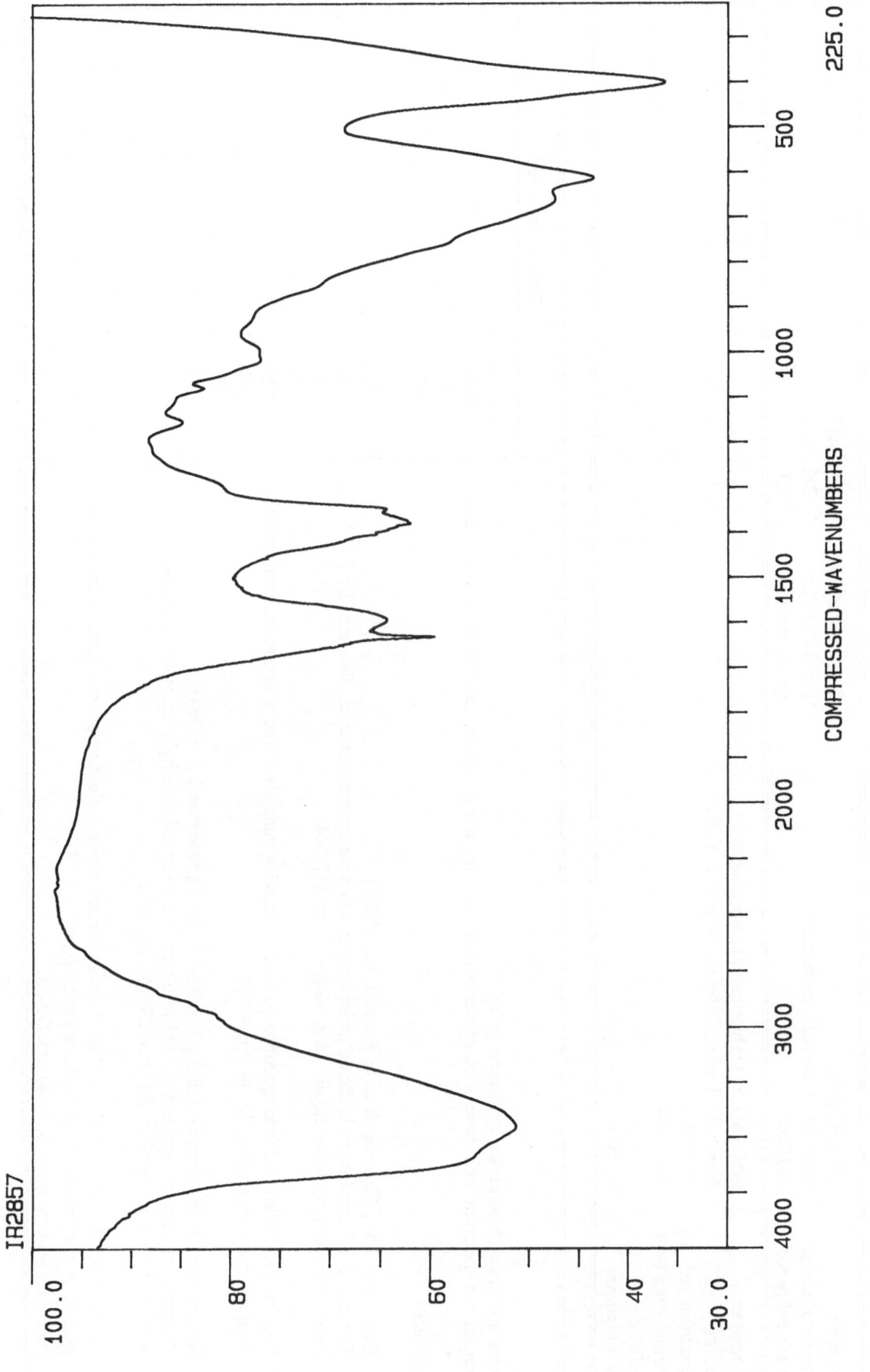

IR2857

% TRANSMITTANCE

COMPRESSED-WAVENUMBERS

DESAUTELSITE

DOLOMITE

Formula:	CaMg(CO$_3$)$_2$
Chemical class:	Anhydrous normal carbonate
Chemical type:	AB(XO$_3$)$_2$

Crystal system:	Trigonal
Mineral group:	Dolomite
Space group:	R$\bar{3}$

Specimen:	BM 1947,52 White rhombs on quartz.
Source:	North Pool mine, Illogan, Cornwall. U.K.
Spectrum ref. No.:	IR2662
Sample medium:	KBr disk
XRD:	
Composition:	Ca:Mg = 1:1 pure

Peak Table cm^{-1}

3019
2896
2628
2528
1821
1441
1090
882
853
729
369
323
262

Notes

Forms a series with **ankerite** and **kutnohorite**.
Compare the spectrum with those of other members of the dolomite group i.e. ankerite & kutnohorite.

References:

1. Dubrawski J.V., Channon A.L. & Warne S.S.J. (1989)
   The effects of substitution in the dolomite ferroan dolomite ankerite series as illustrated by FTIR.
   Neues Jahrbuch für Mineralogie. Monatshefte, **8**, pp.337-344.

2. Rao Yuxue. (1986) Infrared spectroscopy used to identify minerals of the dolomite ankerite series.
   Geology and Prospecting, **22**, (4), pp.41-42.

3. Rakcheev A.D. (Rakcheyev A.D.)., Ragab M.A. & Ventslovaite E.I. (1984)
   Diagnosis of calcium, magnesium, and iron carbonates according to light absorption spectra.
   Moscow University Geology Bulletin, **39**, (6), pp.66-71.

4. Farmer V.C. & Warne S.S.J. (1978) Infrared spectroscopic evaluation of iron contents and excess calcium in minerals of the dolomite ankerite series.
   American Mineralogist, **63** (7,8), pp.779-781.

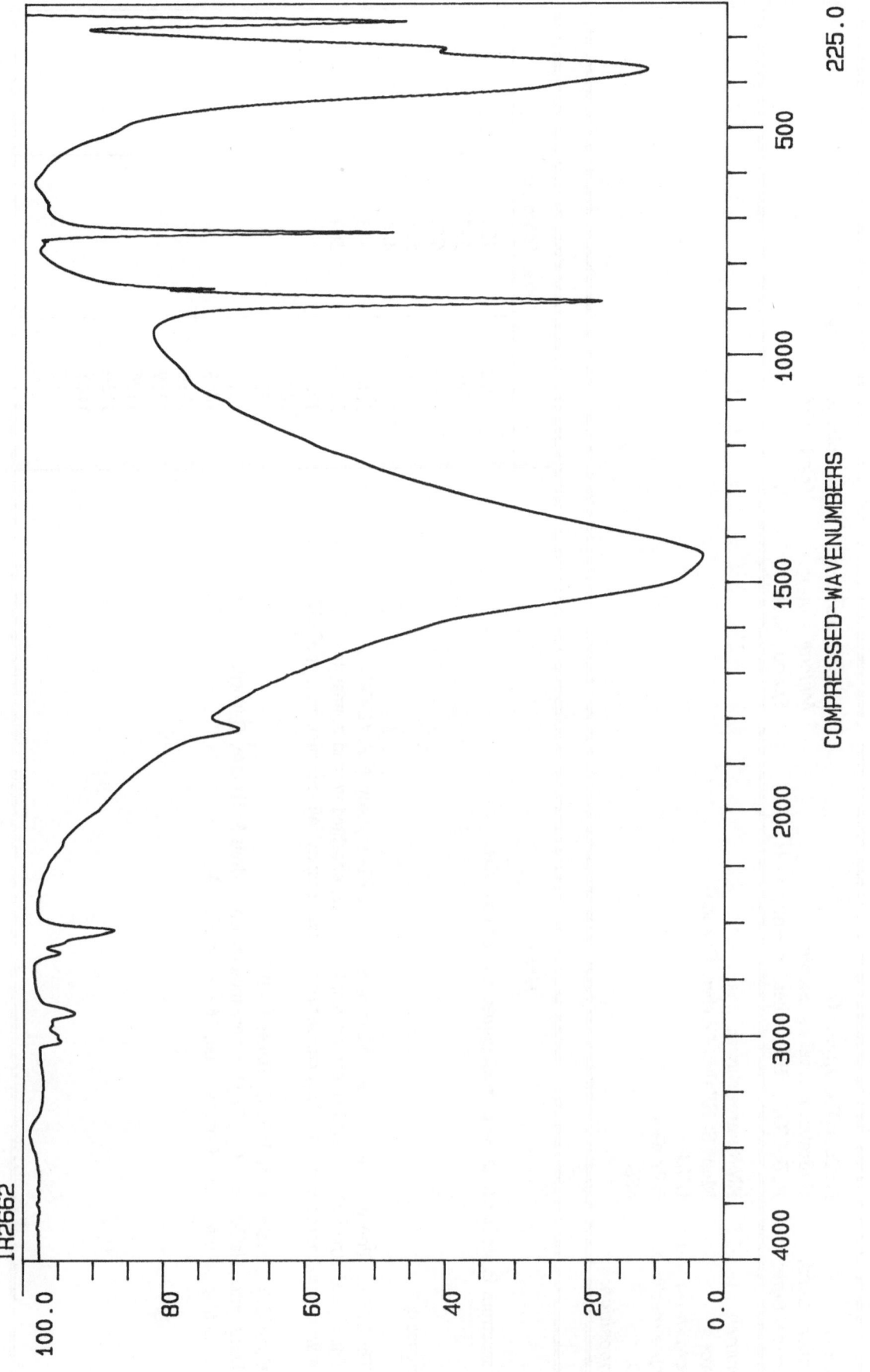

IR2662

% TRANSMITTANCE

100.0 80 60 40 20 0.0

4000 3000 2000 1500 1000 500 225.0

COMPRESSED-WAVENUMBERS

DOLOMITE

DONNAYITE-(Y)

Formula:	NaCaSr$_3$Y(CO$_3$)$_6$·3H$_2$O
Chemical class:	Hydrated normal carbonate
Chemical type:	A$_m$B$_n$(XO$_3$)$_p$·xH$_2$O where (m+n):p = 1:1

Crystal system:	Triclinic, pseudotrigonal
Mineral group:	Mckelveyite
Space group:	P1

Specimen:	RMS unregistered.
Source:	Mont St Hilaire, Quebec, Canada.
Spectrum ref. no.:	IR2919
Sample medium:	KBr disk
XRD:	4556
Composition:	

Peak Table cm^{-1}

3418	**720**
3280	**697**
2956	**632**
2928	521
2857	464
2583	**413**
2428	**284**
1733	
1684	
1523	
1476	
1445	
1396	
1359	
1196	
1150	
1061	
855	

Notes

The spectrum is similar to those of **welloganite** and **mckelveyite**.

References:

1. Thi T.T.L., Pobedimskaya Ye. A., Nadezhina T.N. & Khomyakov A.P. (1984)
   The crystal structures of alkaline carbonates; barentsite; bonshtedtite and donnayite.
   Acta Crystallographica, (A): Foundations of Crystallography, **40** (Supplement), p.C257.

2. Chao G.Y., Mainwaring P.R. & Baker J. (1978)
   Donnayite, NaCaSr$_3$Y(CO$_3$)$_6$·3H$_2$O, a new mineral from Mont St Hilaire, Quebec.
   Canadian Mineralogist, Donnay issue, **16** (3), pp.335-340.

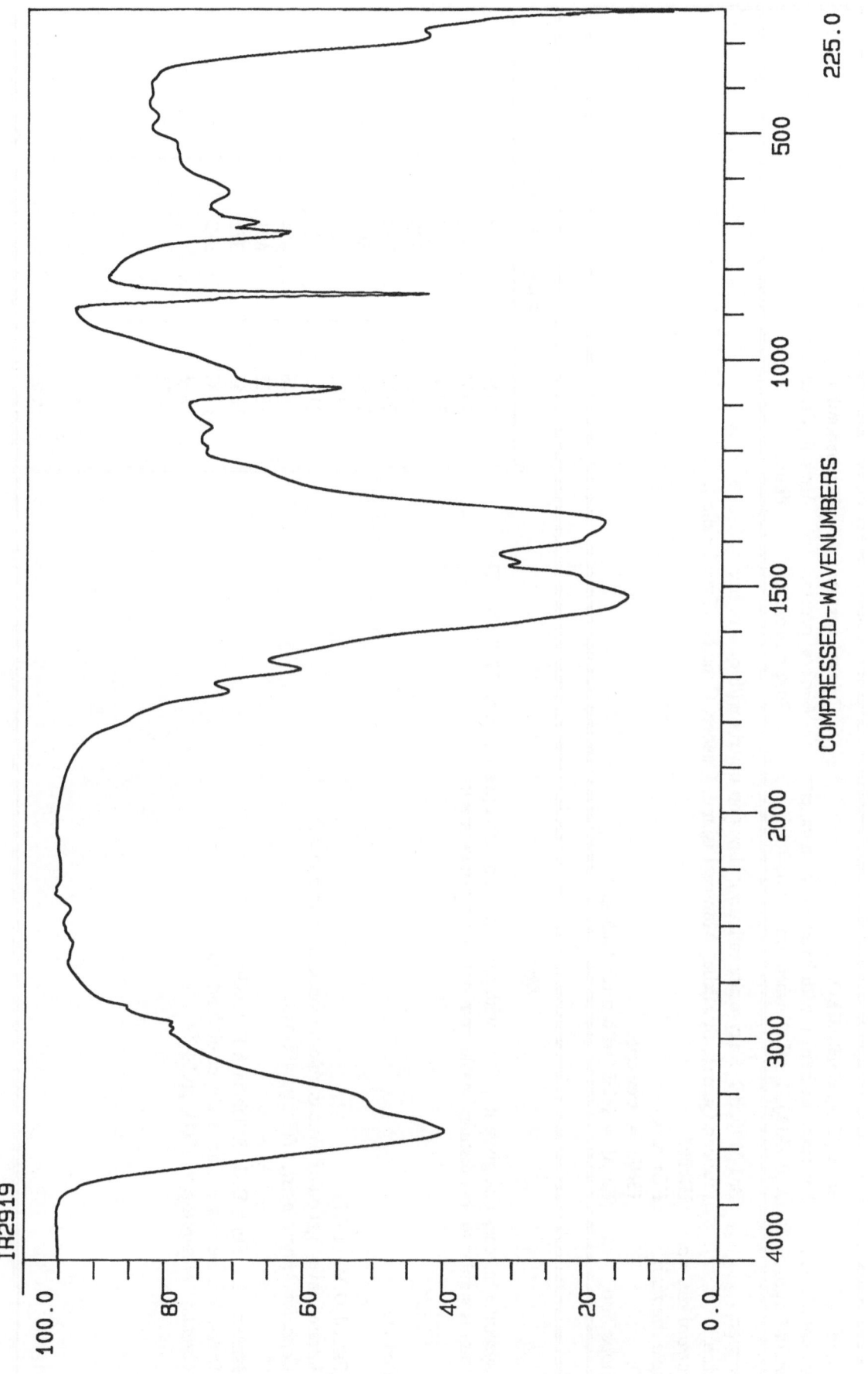

IR2919

% TRANSMITTANCE

100.0

80

60

40

20

0.0

4000

3000

2000

1500

1000

500

225.0

COMPRESSED-WAVENUMBERS

DONNAYITE- (Y)

DRESSERITE

Crystal system:	Orthorhombic
Mineral group:	Alumohydrocalcite
Space group:	Pbnm

Formula:	$Ba_2Al_4(CO_3)_4(OH)_8 \cdot 3H_2O$
Chemical class:	Hydrated carbonate with hydroxyl or halogen
Chemical type:	$A_mB_n(XO_3)_pZ_q \cdot xH_2O$ where $(m+n):p = 3:2$

Specimen	BM 1970,200 Silky white radiating fibres on matrix with quartz etc.
Source:	Francon Quarry, St Michel, Montreal Island, Quebec, Canada. (Type locality).
Spectrum ref. no.:	IR2803
Sample medium:	KBr disk
XRD:	15672 = dresserite
Composition:	Ba:Al = 1:2:3 with trace Na,Mn,Sr.

Peak Table cm⁻¹

3629	859
3485	841
3226	799
2962	753
2923	732
2851	669
2585	567
2132	537
1848	464
1811	401
1641	371
1542	312
1505	
1453	
1376	
1171	
1090	
1041	
954	

Notes

The spectrum matches that given in ref. 1, with the addition of extra peaks around 2900 cm⁻¹. The spectrum is similar to, but distinguishable from that of **strontiodresserite**.

References:

1. Farrell D.M. (1977)
    Infrared investigation of basic double carbonate hydrate minerals.
    Canadian Mineralogist, 15(3), pp.408-413.

2. Jambor J.L., Fong D.G. & Sabina A.P. (1969)
    Dresserite, the new barium analogue of dundasite.
    Canadian Mineralogist, 10(1), pp.84-89.

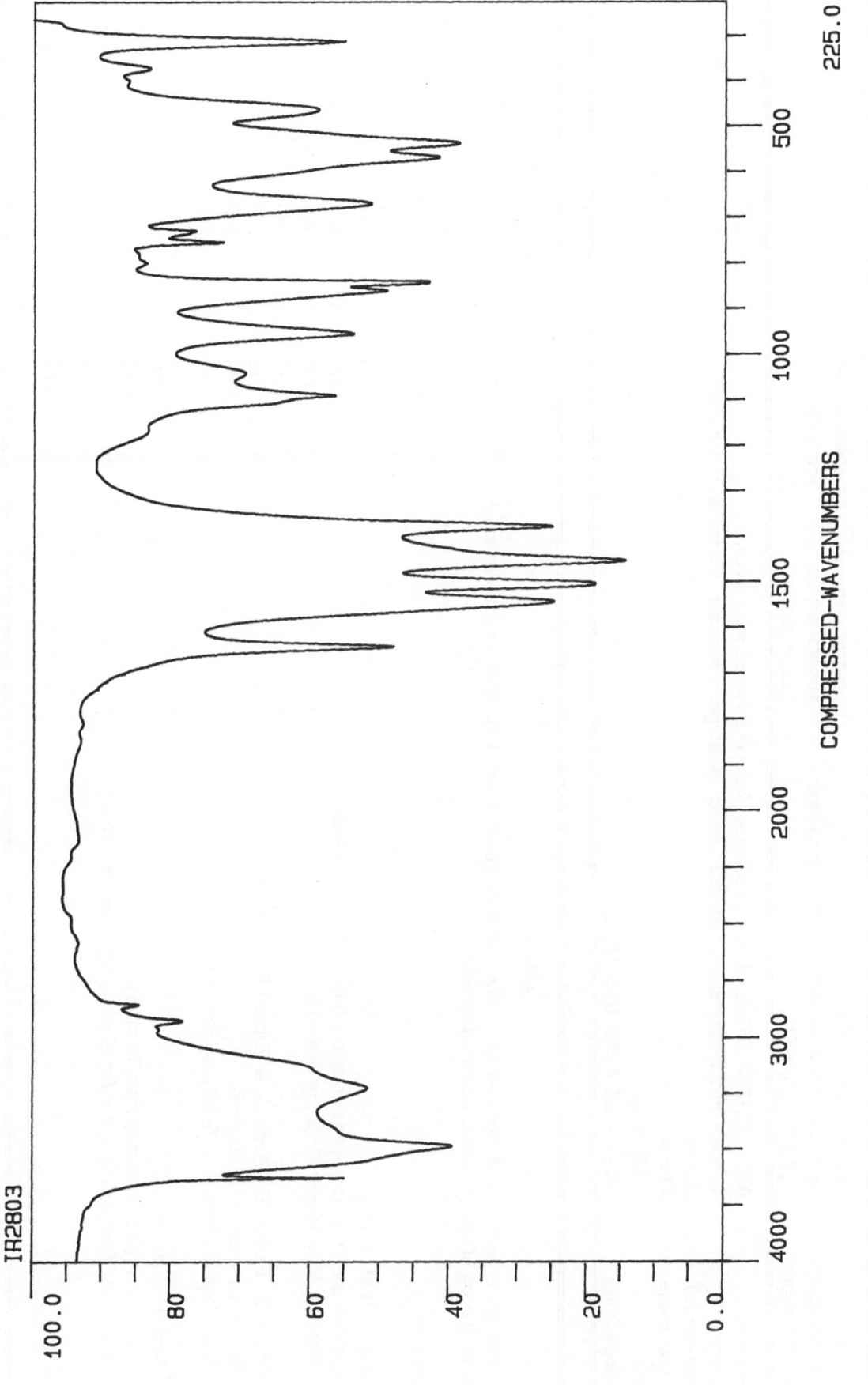

IR2803

% TRANSMITTANCE

100.0　80　60　40　20　0.0

4000　3000　2000　1500　1000　500　225.0

COMPRESSED-WAVENUMBERS

DRESSERITE

DUNDASITE

Formula: $PbAl_2(CO_3)_2(OH)_4 \cdot H_2O$
Chemical class: Hydrated carbonate with hydroxyl or halogen
Chemical type: $A_m B_n (XO_3)_p Z_q \cdot xH_2O$

Crystal system: Orthorhombic
Mineral group: Dundasite
Space group: Pbnm

Specimen: BM 1927,1814 White, fibrous, radiating, botryoidal crust with crocoite.
Source: Adelaide Proprietary mine, Dundas, County Montagu, Tasmania, Australia. (Type locality).
Spectrum ref. no.: IR2795
Sample medium: KBr disk
XRD: 7958F (std)
Composition: Pb:Al = 1:1·4 with trace Si,Fe.

Peak Table cm^{-1}

3596	844		
3450	825		
3076	750		
2926	727		
2858	671		
2484	576		
2277	542		
2197	479		
2115	446		
1810	386		
1642	323		
1523	300		
1506			
1400			
1100			
967			
925			
885			

Notes

The spectrum is close to that shown in ref.1, where peak assignments are given, and comparisons made with the spectra of **dresserite** and **strontiodresserite**.

References:

1. Farrell D.M. (1977)
   Infrared investigation of basic double carbonate hydrate minerals.
   Canadian Mineralogist, 15(3), pp.408-413.

2. Cocco G., Fanfani L., Nunzi A. & Zanazzi P.F. (1972)
   The crystal structure of dundasite.
   Mineralogical Magazine, 38(297), pp.564-569.

3. Ford R. J. (1967)
   A new analysis of dundasite from Tasmania.
   Papers and Proceedings of the Royal Society of Tasmania, 101, p.9.

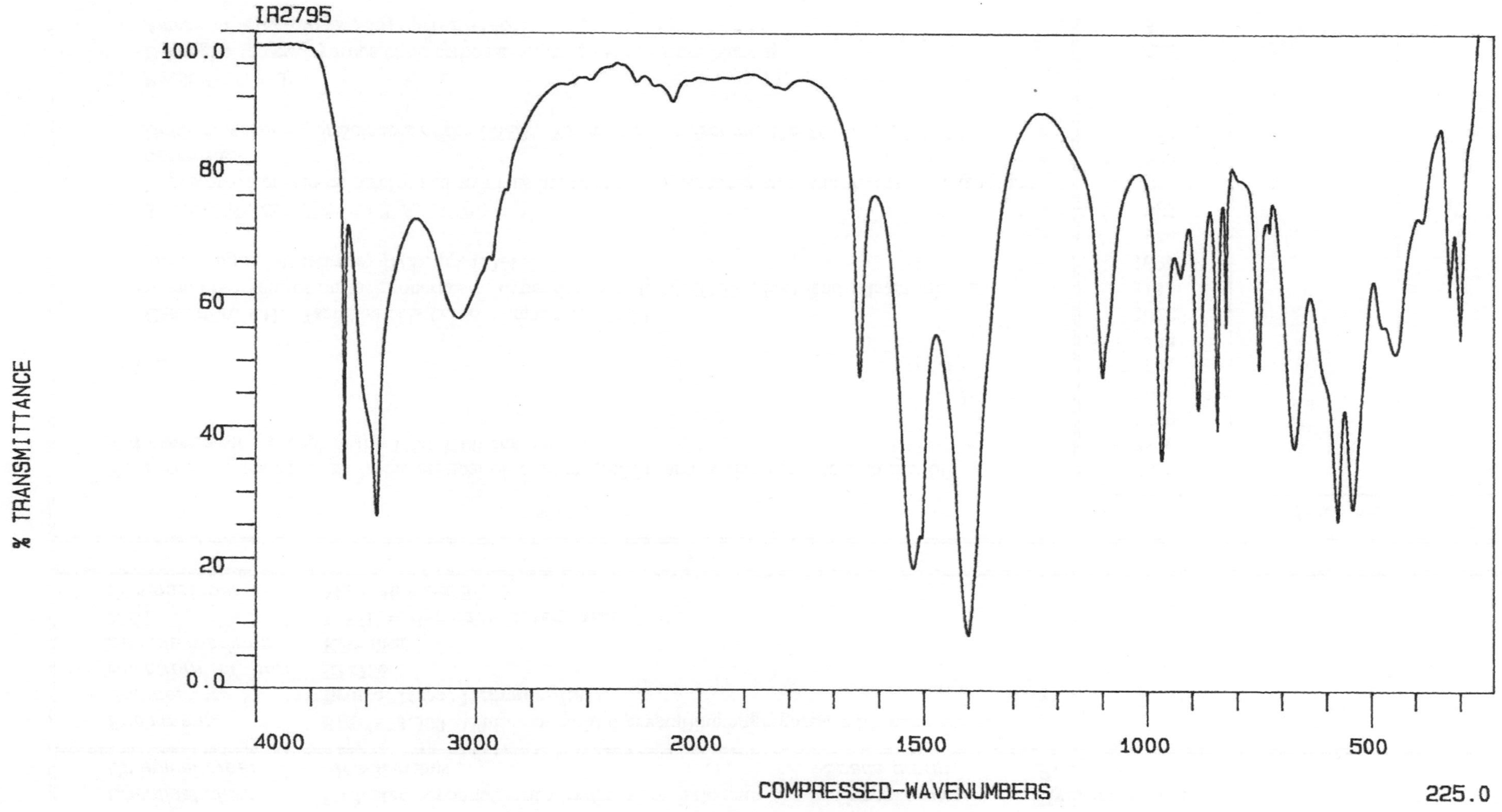

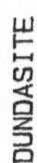

IR2795

% TRANSMITTANCE

100.0 80 60 40 20 0.0

4000 3000 2000 1500 1000 500 225.0

COMPRESSED-WAVENUMBERS

DUNDASITE

DYPINGITE

Formula:	$Mg_5(CO_3)_4(OH)_2 \cdot 5H_2O$
Chemical class:	Hydrated carbonate with hydroxyl or halogen
Chemical type:	Miscellaneous

Crystal system:	Monoclinic ?
Mineral group:	Hydromagnesite
Space group:	P1 ?

Specimen:	BM 1978,500 White botryoidal crystalline aggregates, with canavesite.
Source:	Brosso, Ivrea, Piemonte, Italy.
Spectrum ref. no.:	IR2786
Sample medium:	KBr disk
XRD:	7857F = dypingite or very near
Composition:	Mg with trace S.

Peak Table cm⁻¹

3650
3510
3443
2929
1601
1484
1428
1113
1097
944
883
855
799
718
665
596
425
381

Notes

The spectrum matches that in the original description (ref.3), and is also very close to that of **hydromagnesite**, except in the 3400-3700 cm⁻¹ region.

References:

1. Canterford J.H., Tsambourakis G. & Lambert R. (1984)
   Some observations on the properties of dypingite, $Mg_5(CO_3)_4 (OH)_2 \cdot 5H_2O$, and related minerals.
   Mineralogical Magazine, **48**(3), pp.437-442.

2. Smolin P.P. & Ziborova T.A. (1976)
   Types of water, stoichiometry and relations between hydromagnesite and other hydrated magnesium carbonates.
   Doklady-Academy of Sciences of the USSR, Earth Sciences Section, **226**(16), pp.130-133.

3. Raade G. (1970)
   Dypingite, a new hydrous basic carbonate of magnesium, from Norway.
   American Mineralogist, **55**, pp.1457-1465.

IR2786

% TRANSMITTANCE

100.0   80   60   40   20.0

4000   3000   2000   1500   1000   500   225.0

COMPRESSED-WAVENUMBERS

DYPINGITE

GASPÉITE

Crystal system:	Trigonal
Mineral group:	Calcite
Space group:	R$\bar{3}$c

Formula:	(Ni,Mg,Fe)CO$_3$ (see notes)
Chemical class:	Anhydrous normal carbonate
Chemical type:	A(XO$_3$)
Specimen:	BM 1985,497 Pale green/yellow coating with kambaldaite.
Source:	Otter Shoot, Kambalda, Kalgoorlie, Western Australia.
Spectrum ref. no.:	IR2826
Sample medium:	KBr disk
XRD:	6705F (std)
Composition:	Ni:Mg = 1: <0·1 with trace Fe

Peak Table cm^{-1}

3511		
3483		
3295		
2925		
2859		
2502		
1825		
1438		
1086		
965		
871		
753		
531		
374		
240?		

Notes

Forms a series with magnesite. Specimens with high nickel and low iron are rare, many 'gaspéites' are magnesian gaspéite or nickeloan magnesite. This specimen is close to the ideal formula NiCO$_3$. Compare the spectrum with that in Suhner (5-29 A), and original data (ref.2), which relate to zincian and magnesian species.

References:

1. White W.B. (1974)
   The carbonate minerals.
   In: Farmer (Ed) *The Infrared Spectra of Minerals.*
   Mineralogical Society of London, Monograph No. 4, pp.227-284.

2. Kohls D.W. & Rodda J.L. (1966)
   Gaspéite, (Ni,Mg,Fe)(CO$_3$) a new carbonate from the Gaspé Peninsula, Quebec.
   American Mineralogist, **51**(5,6), pp.677-684.

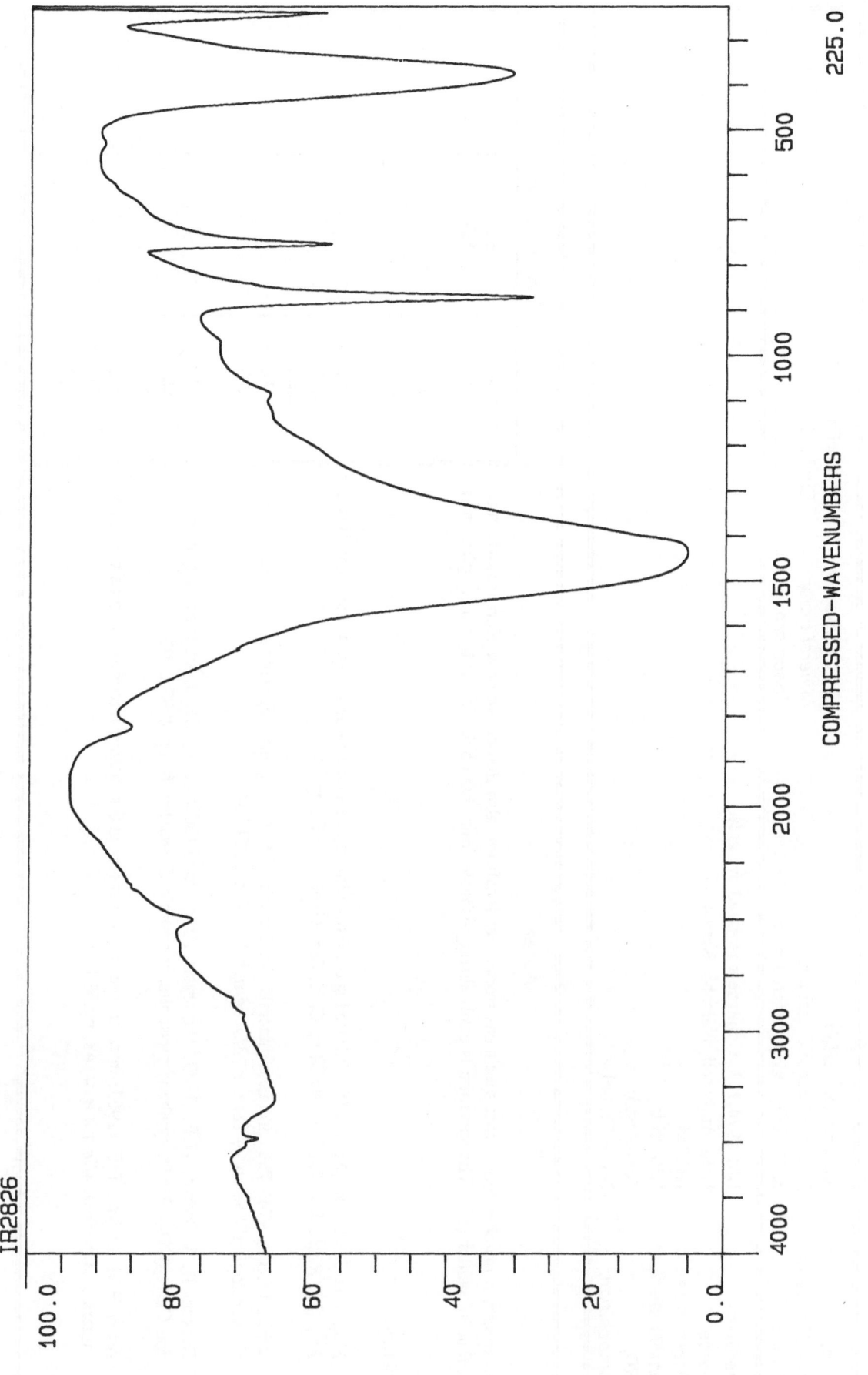

IR2826

% TRANSMITTANCE

COMPRESSED-WAVENUMBERS

GASPEITE

225.0   500   1000   1500   2000   3000   4000

100.0   80   60   40   20   0.0

GAYLUSSITE

Crystal system:	Monoclinic
Mineral group:	Gaylussite
Space group:	C2/c or Cc

Formula:	$Na_2Ca(CO_3)_2 \cdot 5H_2O$
Chemical class:	Hydrated normal carbonate
Chemical type:	$A_m B_n(XO_3)_p \cdot xH_2O$ where (m+n):p > 1:1
Specimen:	BM 1974,217 Colourless isolated crystals.
Source:	Lake Amboseli, Nairobi, Kenya.
Spectrum ref. no.:	IR2783
Sample medium:	KBr disk
XRD:	7854F (std)
Composition:	Na & Ca only

Peak Table cm⁻¹

3345	523
2966	267
2499	
2461	
2393	
1787	
1662	
1617	
1444	
1414	
1070	
898	
876	
805	
720	
693	
653	
557	

Notes

The spectrum matches those from specimens from other localities, also that of Adler & Kerr (ref.4), but not that in Sadtler (80). The spectrum is easily distinguishable from that of the lower hydrate, **pirssonite.**

References:

1. Maglione G. & Carn M. (1975) Infrared spectra of saline and silicate minerals from the Chad Basin. *Fr., Off. Rech. Sci. Tech. Outre Mer, Cah., Ser. Geol.* 7, (1), pp. 3-9.

2. White W.B. (1974) The carbonate minerals. *In:* Farmer (Ed) *The Infrared Spectra of Minerals. Mineralogical Society of London, Monograph No. 4,* pp.227-284.

3. Dickens B. & Brown. W.E. (1969) The crystal structure of $CaNa_2(CO_3)_2 \cdot 5H_2O$, synthetic gaylussite, and $CaNa_2(CO_3)_2 \cdot 2H_2O$, synthetic pirssonite. *Inorganic Chemistry,* **8,** pp.2093-2103

4. Adler H.H. & Kerr P.F. (1963) Infrared spectra, symmetry and structure relations of some carbonate minerals. *American Mineralogist,* **48,** pp.839-853.

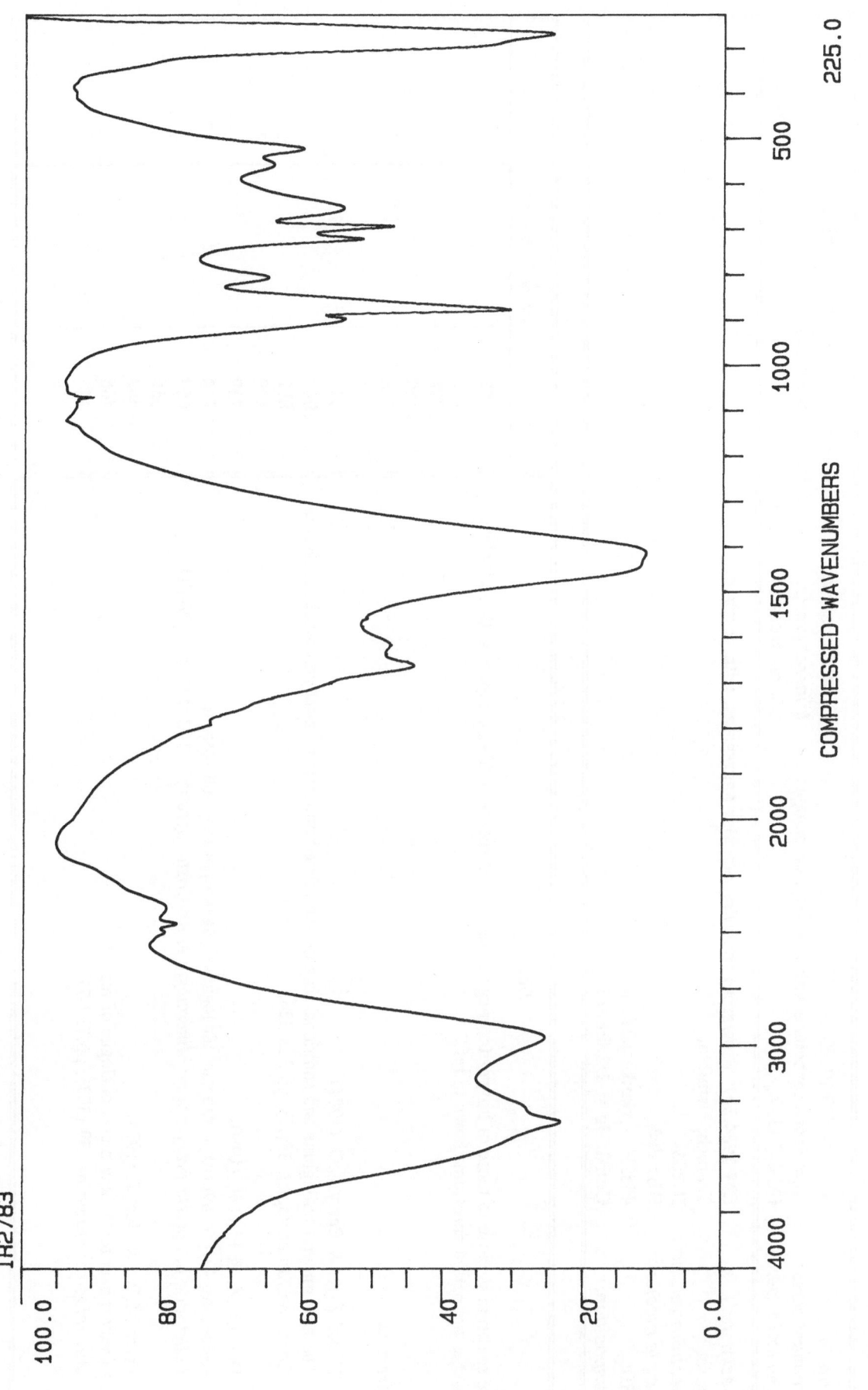

IR2783

% TRANSMITTANCE

COMPRESSED-WAVENUMBERS

GAYLUSSITE

GLAUKOSPHAERITE

Formula:	$(Cu,Ni)_2(CO_3)(OH)_2$
Chemical class:	Anhydrous carbonate with hydroxyl or halogen
Chemical type:	$(AB)_2(XO_3)Z_q$
Crystal system:	Monoclinic, pseudo-orthorhombic.
Mineral group:	Malachite (rosasite)
Space group:	?

Specimen:	BM 1984,381 Blue/green radiating acicular aggregates, with rosasite.
Source:	Tsumeb, Namibia.
Spectrum ref. no.:	IR2869
Sample medium:	KBr disk
XRD:	4082F = glaukosphaerite
Composition:	Cu:Ni:Mg = 1·2:0·8:0·1

Peak Table cm⁻¹

3497
3247
1528
1420
1384
1174
1099
1049
852
828
739
704
669
556
464
406
327
274

Notes

The spectrum is close to those of kolwezite, mcguinnessite and rosasite. Also compare with malachite.
Matches the partial spectrum shown in ref.2.

References:

1. Nickel E.H. & Berry L.G. (1981)
    The new mineral nullaginite and additional data on the related minerals rosasite and glaukosphaerite.
    Canadian Mineralogist, **19**,(2), pp.315-324.

2. Deliens M. & Piret P. (1980)
    Kolwezite, Cu-Co hydroxycarbonate, analogue of glaukosphaerite and rosasite.
    Bulletin de la Société Française de Minéralogie et de Cristallographie, **103**, (2), pp.179-184.

3. Pryce M.W. & Just J. (1974)
    Glaukosphaerite: A new nickel analogue of rosasite.
    Mineralogical Magazine, **39**,(307), pp.737-743.

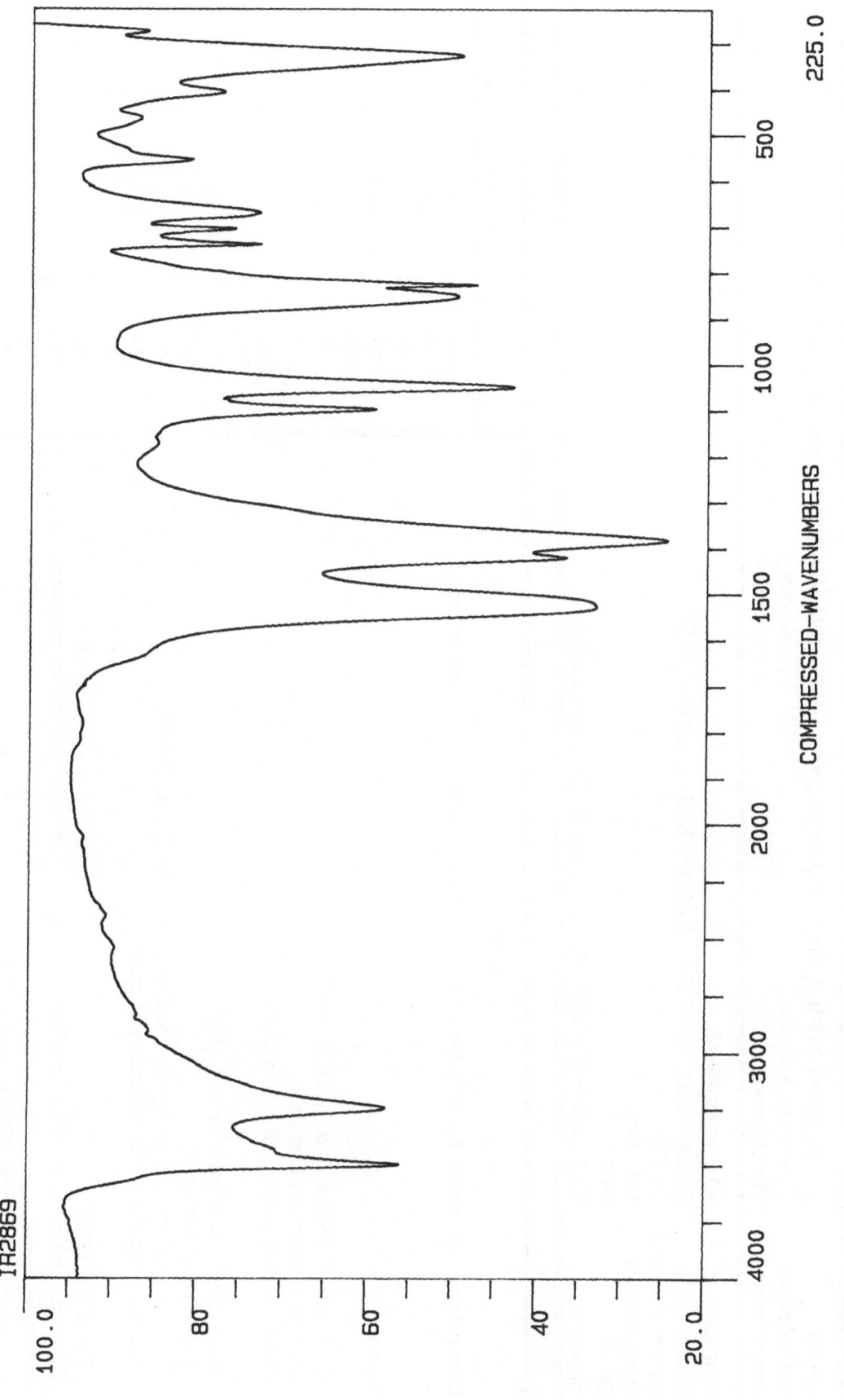

IR2869

% TRANSMITTANCE

COMPRESSED-WAVENUMBERS

GLAUKOSPHAERITE

HARKERITE

Formula:	$Ca_{24}Mg_8[AlSi_4(O,OH)_{16}]_2(BO_3)_8(CO_3)_8(H_2O,Cl)$ *
Chemical class:	Compound carbonate
Chemical type:	Miscellaneous

Crystal system:	Trigonal, pseudo-cubic
Mineral group:	
Space group:	R3m

Specimen:	RMS unregistered.
Source:	Camas Malag, Isle of Skye, Highland Region, Scotland, U.K.
Spectrum ref. no.:	IR2926
Sample medium:	KBr disk
XRD:	9997
Composition:	Ca:Mg:Al:Si:Cl ≈ 12·5:1:5:0·5 + trace Fe(B not determined)

Peak Table cm⁻¹

3685	612
3435	585
2953	**538**
2928	**455**
2859	403
2594	**319**
2515	
1793	
1734	
1515	
1242	
976	
904	
877	
861	
852	
776	
741	
713	

Notes

* Idealised unit cell content. See ref.1 for a discussion of the structure and unit cell contents.

References:

1. Giuseppetti G., Mazzi F. & Tadini C. (1977)
   The crystal structure of harkerite.
   American Mineralogist, **62**(3,4), pp.263-272.

2. Malinko S.V. & Kuznetsova N.N. (1973)
   A new find of sakhaite.
   Zapiski Vsesoiuznoe Mineralogicheskoe Obshchestvo, **102**(2), pp.164-170.
   (includes comparison with harkerite & IR spectra).

3. Tilley C.E. (1951)
   The zoned contact skarns of the Broadford area, Skye; a study of boron fluorine metasomatism in dolomites.
   Mineralogical Magazine, **29**(214), pp.621-666.

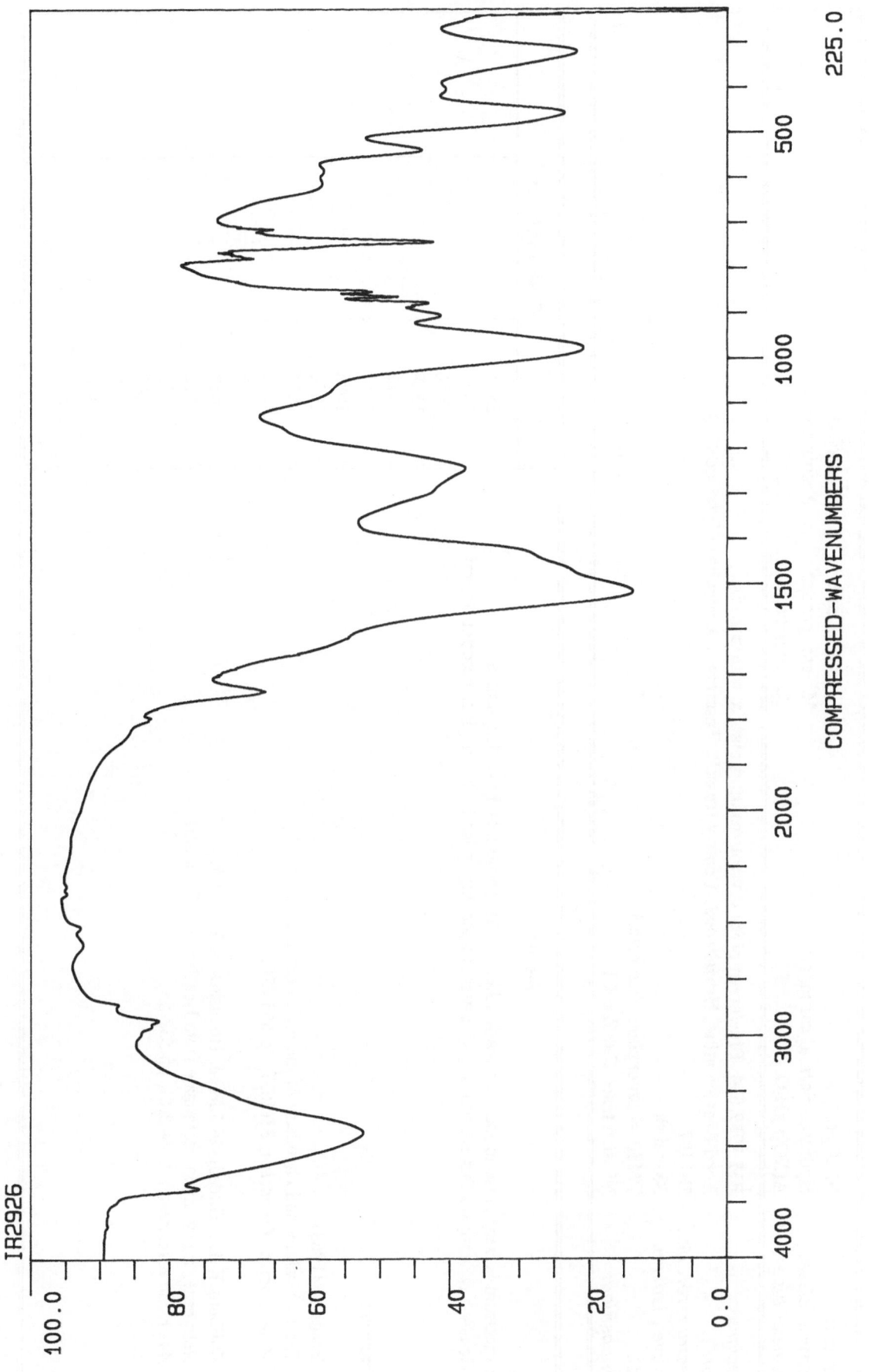

IR2926

% TRANSMITTANCE

COMPRESSED-WAVENUMBERS

HARKERITE

HELLYERITE

Formula:	$NiCO_3 \cdot 6H_2O$	
Chemical class:	Hydrated normal carbonate	
Chemical type:	$A(XO_3) \cdot xH_2O$	
Crystal system:	Monoclinic	
Mineral group:	Nesquehonite	
Space group:	C2/c	

Specimen:	BM 1959,534 Blue/green platy crystals coating matrix with zaratite.
Source:	Lord Brassey mine, Heazelwood, County Russell, Tasmania, Australia. (Type locality).
Spectrum ref. no.:	IR2797
Sample medium:	KBr disk
XRD:	7948F = amorphous (see notes)
Composition:	Ni with trace Cu, Zn, Cl

Peak Table cm⁻¹

3411
2926
2856
1568
1422
1161
1086
1030
837
799
779
680
622
513
463
399
374
325

Notes

The spectrum is very close to that of **zaratite**, i.e. the hellyerite may have dehydrated. Previous x-ray work on this specimen gave patterns matching those in the original description, ref.2.

References:

1. Isaacs T. (1963)
    The mineralogy and chemistry of the nickel carbonates.
    Mineralogical Magazine, **33**(263), pp.663-678.

2. Williams K.L., Threadgold I.M. & Hounslow A.W. (1959)
    Hellyerite, a new nickel carbonate from Heazlewood, Tasmania.
    American Mineralogist, **44**(5,6), pp.533-538.

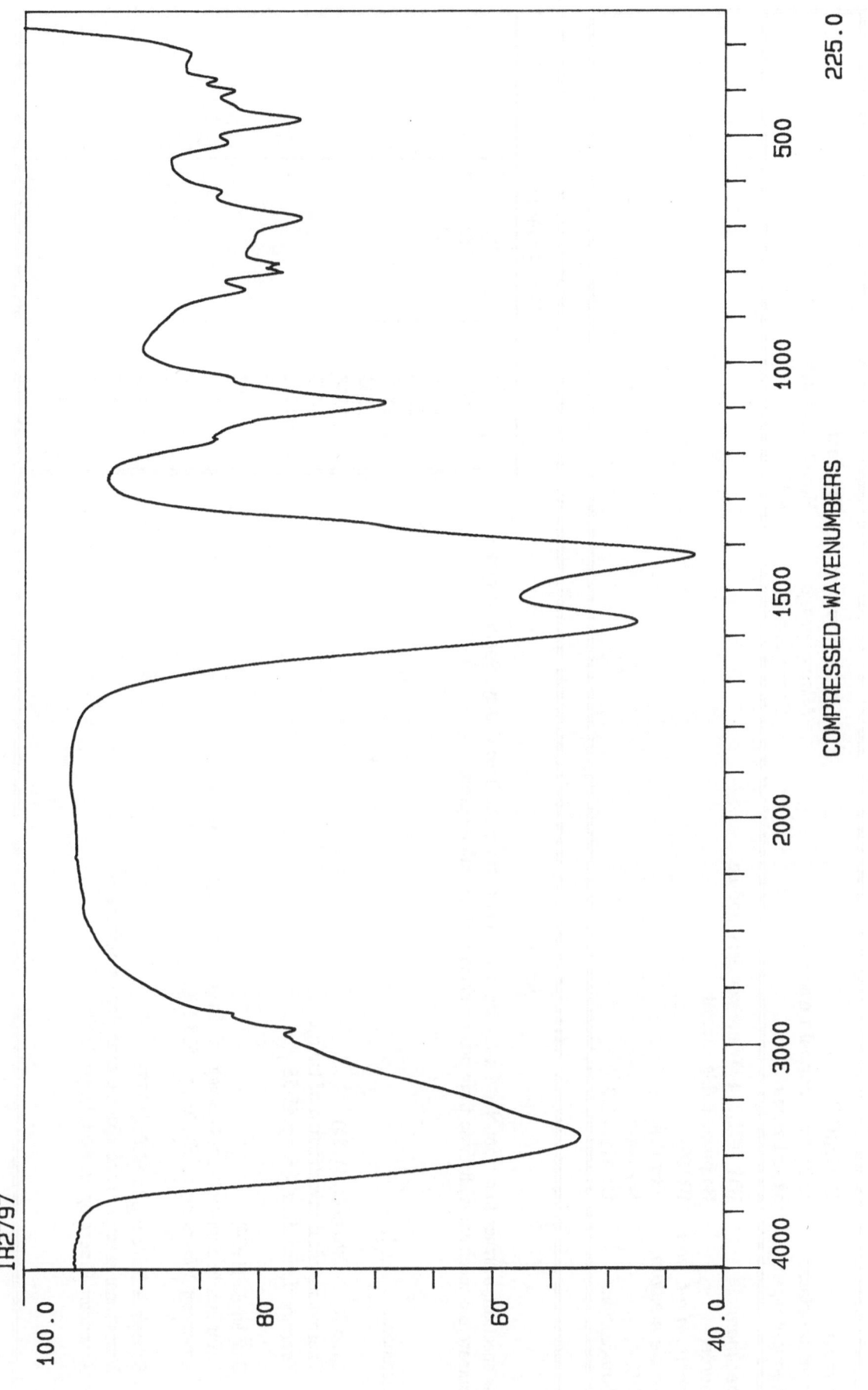

IR2797

% TRANSMITTANCE

100.0

80

60

40.0

4000   3000   2000   1500   1000   500   225.0

COMPRESSED-WAVENUMBERS

HELLYERITE

HUNTITE

Crystal system:	**Trigonal**
Mineral group:	**Huntite (calcite)**
Space group:	**R32**

Formula:	**CaMg₃(CO₃)₄**
Chemical class:	**Anhydrous normal carbonate**
Chemical type:	**Miscellaneous**
Specimen:	**BM 1972,214 White powdery nodule.**
Source:	**Boquira, Bahia, Brazil.**
Spectrum ref. no.:	**IR2669**
Sample medium:	**KBr disk**
XRD:	**See notes**
Composition:	**Ca:Mg ≈ 1:3**

Note: *Formula:* **CaMg$_3$(CO$_3$)$_4$**

Peak Table cm⁻¹

2979	**386**
2901	**283**
2584	249?
2546	
1828	
1543	
1508	
1463	
1442	
113	
891	
887	
870	
851	
744	
697	
667	
449	

Notes

The spectrum matches that from NHM x-ray standard huntite from Tea Tree Gulley, South Australia.
Compare the spectrum with those from other members of the **calcite** group.

References:

1. Orzao R. & Otsuka R. (1985)
   Thermoanalytical investigation of huntite.
   Thermochimica Acta, **86**, pp.45-58.

2. Shayan A. (1984)
   Strontium in huntites from Geelong and Deer Park, Victoria, Australia.
   American Mineralogist, **69**(5,6), pp.528-530.

3. Scheetz B.E. & White W.B. (1977)
   Vibrational spectra of the alkaline earth double carbonates.
   American Mineralogist, **62**(1,2), pp.36-50.

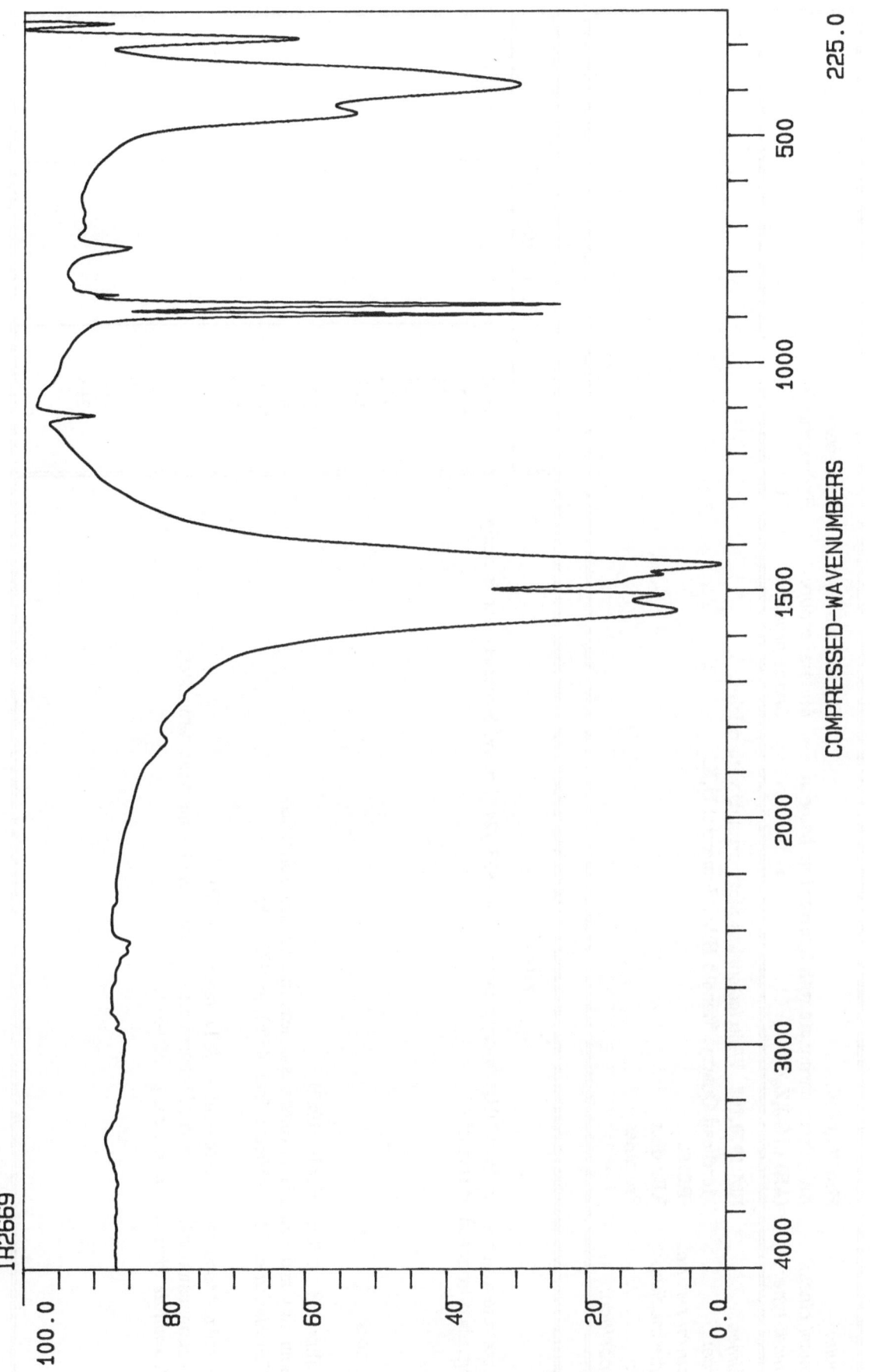

IR2669

% TRANSMITTANCE

COMPRESSED-WAVENUMBERS

4000 3000 2000 1500 1000 500 225.0

100.0 80 60 40 20 0.0

HUNTITE

HYDROCERUSSITE

Formula:	**Pb$_3$(CO$_3$)$_2$(OH)$_2$**
Chemical class:	**Anhydrous carbonate with hydroxyl or halogen**
Chemical type:	**(AB)$_3$(XO$_3$)$_2$Z$_q$**
Crystal system:	**Hexagonal**
Mineral group:	**Hydrocerussite**
Space group:	**P1 ?**

Specimen:	**BM 1970,108 White/colourless platy crystals with calcite.**
Source:	**Merehead Quarry, Mendip Hills, Somerset, U.K.**
Spectrum ref. no.:	**IR2750**
Sample medium:	**KBr disk**
XRD:	**See notes**
Composition:	

Peak Table cm^{-1}

3530
3450
2923
2419
1736
1631
1410
1226
1099
1046
850
781
693
683
620
468
393

Notes

The spectrum matches that from NHM, x-ray standard, BM 1923,724 (Priddy, Somerset), and is easily distinguished from that of **cerussite**.

References:

1. Bilinski H. & Schindler P. (1982)
   Solubility and equilibrium constants of lead in carbonate solutions.
   Geochimica et Cosmochimica Acta, **46**(6), pp.921-928.

2. Bessière-Morandat J., Lorenzelli V. & Lecomte J. (1970)
   Determination and attribution of infrared active vibrations of some basic carbonates.
   Journal de Physique, Paris, **31**, pp.309-312.

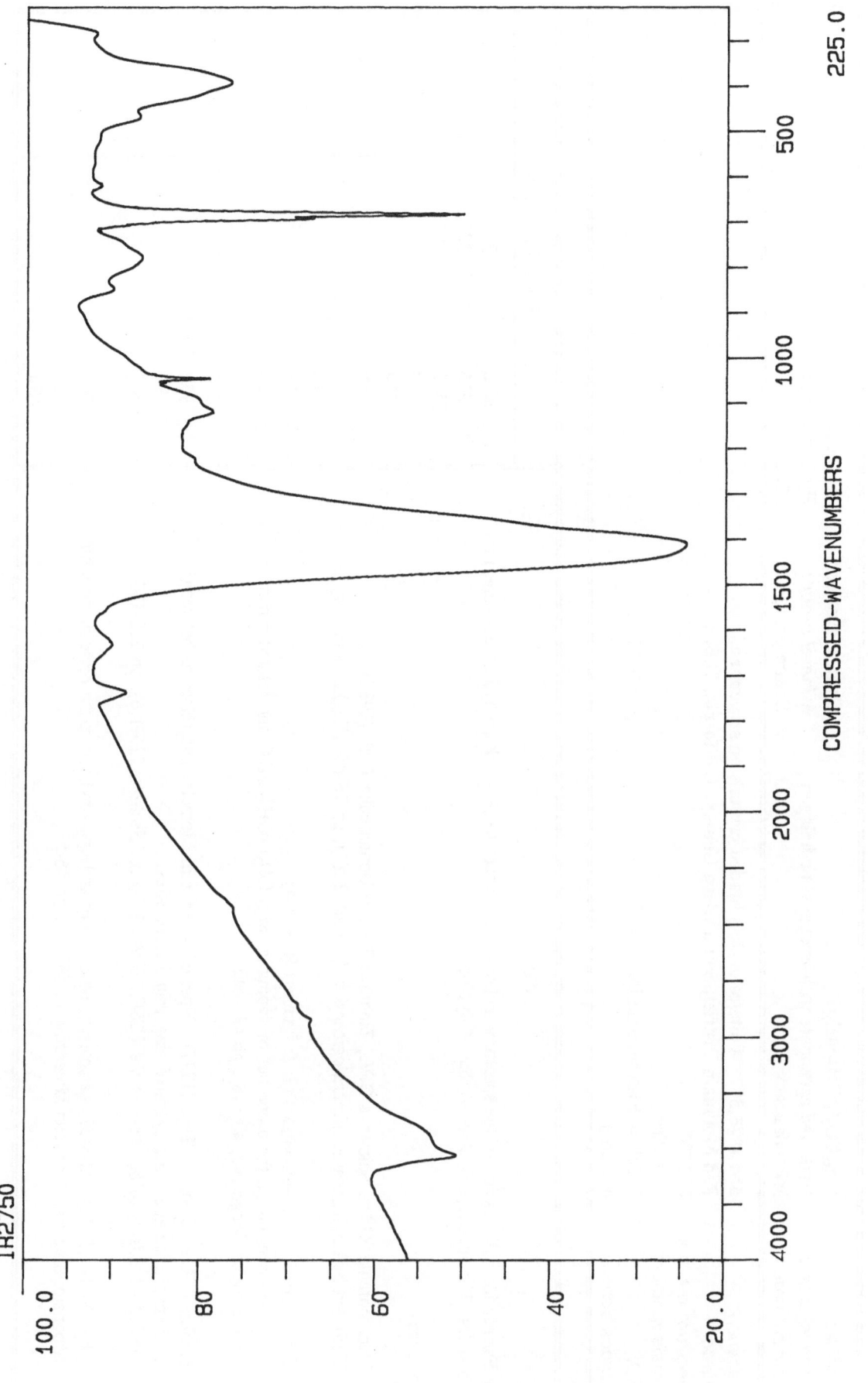

IR2750

% TRANSMITTANCE

100.0
80
60
40
20.0

4000
3000
2000
1500
1000
500
225.0

COMPRESSED-WAVENUMBERS

HYDROCERUSSITE

HYDROMAGNESITE

Formula:	$Mg_5(CO_3)_4(OH)_2 \cdot 4H_2O$
Chemical class:	Hydrated carbonate with hydroxyl or halogen
Chemical type:	Miscellaneous
Crystal system:	Monoclinic
Mineral group:	Hydromagnesite
Space group:	$P2_1/c$

Specimen:	BM 1975,597 White/colourless bladed crystals on serpentine.
Source:	Red Mountain District, Santa Clara County, California, U.S.A.
Spectrum ref. no.:	IR2770
Sample medium:	KBr disk
XRD:	19248 = hydromagnesite
Composition:	Mg only

Peak Table cm⁻¹

3650	**379**
3516	**337**
3452	250?
3237	
2994	
2588	
2541	
1483	
1428	
1120	
1109	
884	
853	
786	
747	
714	
596	
472	
436	

Notes

See Farmer for a discussion of the spectrum and other references to IR work on hydrated magnesium carbonates. The spectrum is close to that of **dypingite**.

References:

1. Nechiporenko G.O., Sokolova G.V., Ziborova T.A. & Bondarenko G.P. (1988) Hydrated hydromagnesite. *Mineralogicheskiy Zhurnal*, **10**(1), pp.78-85. (English summary).

2. Canterford J.H., Tsambourakis G. & Lambert R. (1984) Some observations on the properties of dypingite, $Mg_5(CO_3)_4(OH)_2 \cdot 5H_2O$ and related minerals. *Mineralogical Magazine*, **48**(348), pp.437-442.

3. Smolin P.P. & Ziborova T.A. (1977) Types of water, stoichiometry and relations between hydromagnesite and other hydrated magnesium carbonates. *Doklady-Academy of Sciences of the USSR, Earth Sciences Section*, **226**(1,6), pp.130-133.

4. White W.B. (1971) Infrared characterization of water and hydroxyl ion in the basic magnesium carbonate minerals. *American Mineralogist*, **56**(1,2), pp.46-53.

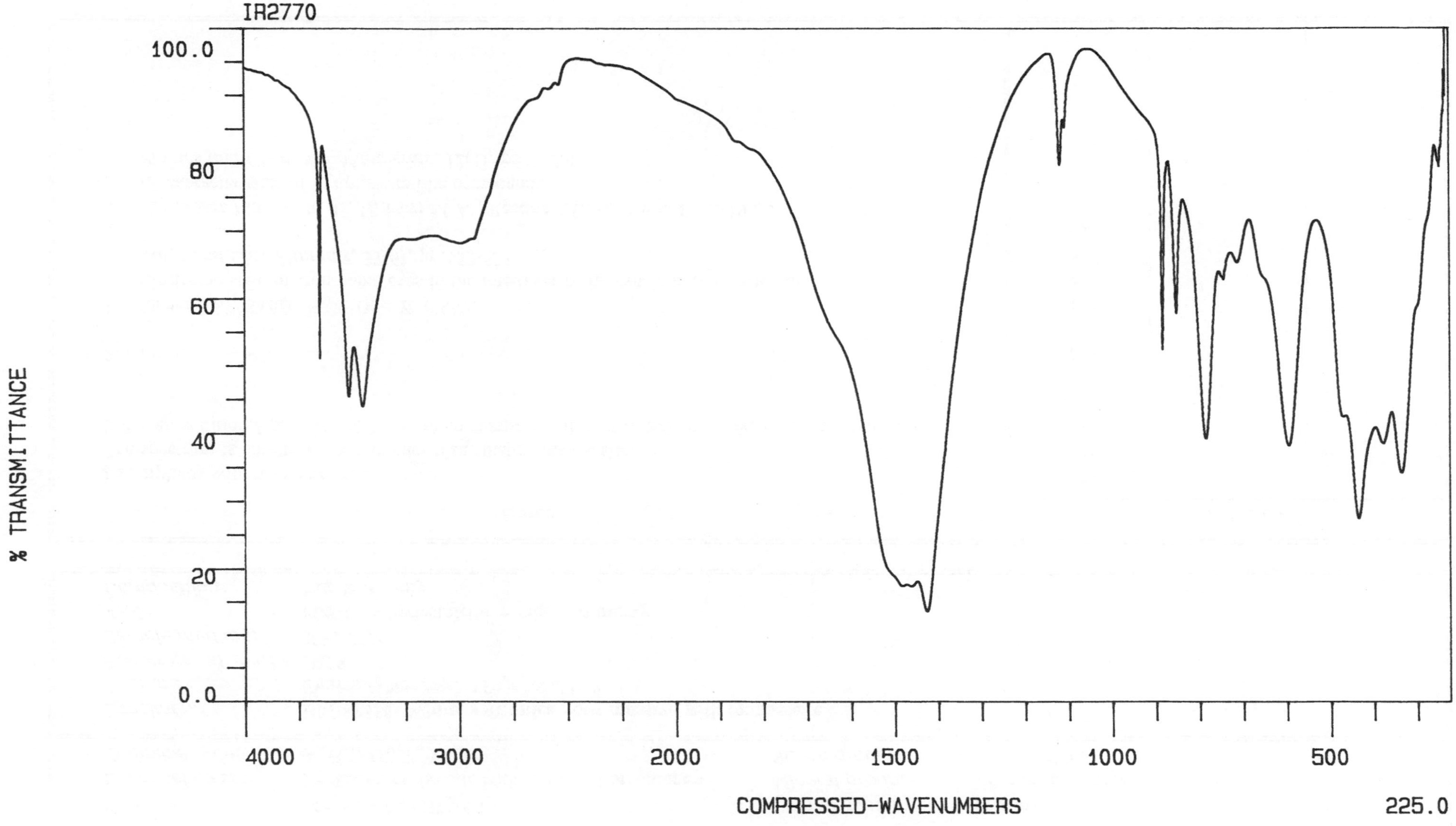

IR2770

% TRANSMITTANCE

100.0　　80　　60　　40　　20　　0.0

4000　　3000　　2000　　1500　　1000　　500　　225.0

COMPRESSED-WAVENUMBERS

HYDROMAGNESITE

HYDROTALCITE

Formula:	$Mg_6Al_2(CO_3)OH_{16} \cdot 4H_2O$
Chemical class:	Hydrated carbonate with hydroxyl or halogen
Chemical type:	$A_mB_n(XO_3)_pZ_q \cdot xH_2O$
Crystal system:	Trigonal
Mineral group:	Sjögrenite (hydrotalcite)
Space group:	$R\bar{3}m$ or R3m

Specimen:	BM 89358 White, soft, micaceous massive with manasseïte.
Source :	Snarum, Norway. (Type locality ?).
Spectrum ref. no.:	IR2853
Sample medium:	KBr disk
XRD:	8067F = hydrotalcite + slight impurity
Composition:	Mg & Al only

Peak Table cm^{-1}

3573		
3510		
2431		
1653		
1632		
1583		
1372		
	855	
	664	
	557	
	414	

Notes

Dimorphous with manasseïte.

The spectrum is similar to, but simpler than, that of manasseïte.

It has an additional peak at 1632 cm^{-1} in comparison with that in Suhner (5-66 A) hydrotalcite.

References:

1. Idemura S., Suzuki E. & Ono Y. (1989)
   Electronic state of iron complexes in the interlayer of hydrotalcite like materials.
   Clays and Clay Minerals, 37(6), pp.553-557.

2. Hernandez Moreno M.J., Ulibarri M.A., Rendon J.L. & Serna C.J. (1985)
   IR characteristics of hydrotalcite-like compounds.
   Physics and Chemistry of Minerals, 12(1), pp.34-38.

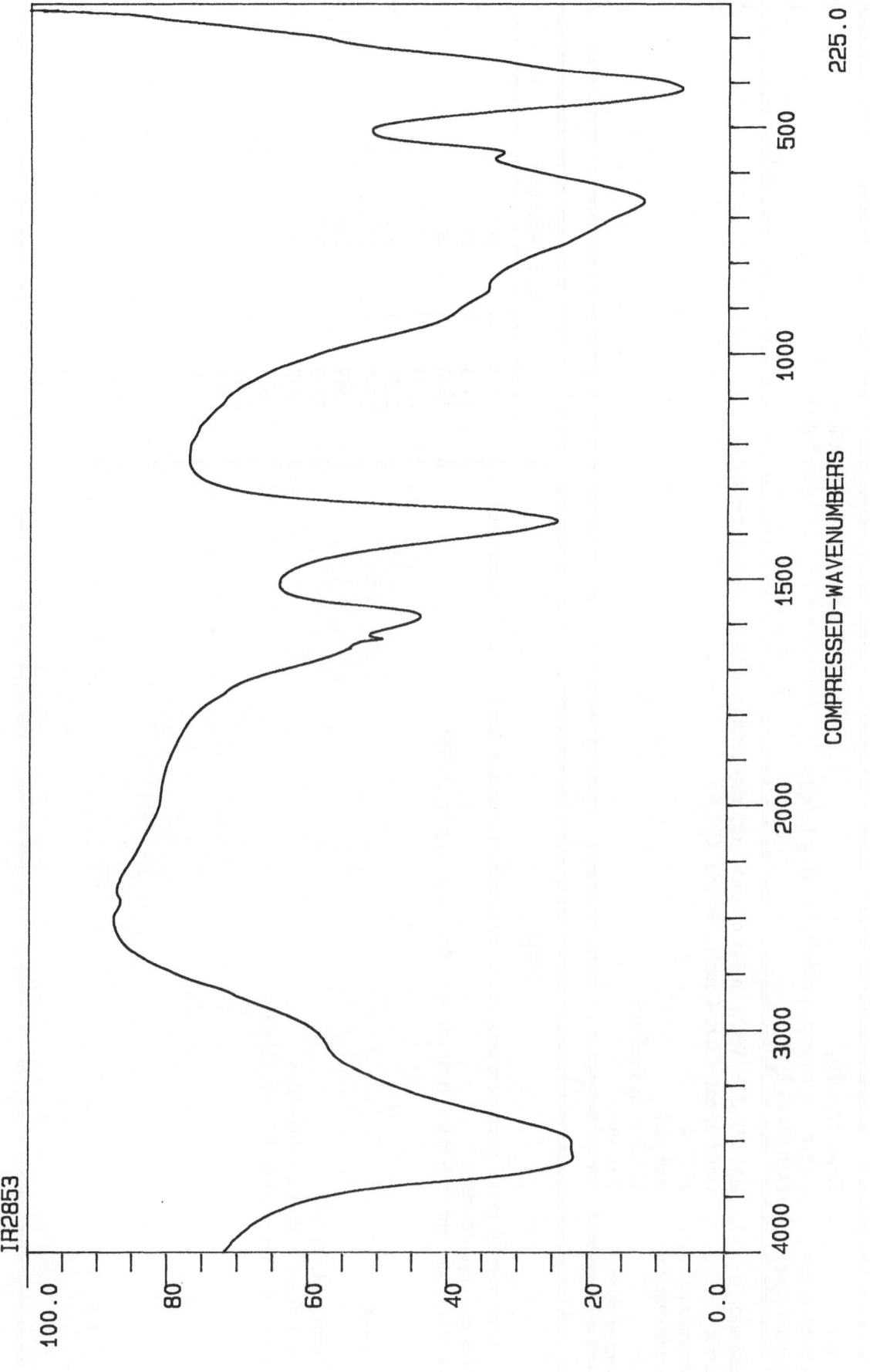

IR2853

% TRANSMITTANCE

100.0   80   60   40   20   0.0

4000   3000   2000   1500   1000   500   225.0

COMPRESSED-WAVENUMBERS

HYDROTALCITE

HYDROZINCITE

Formula:	$Zn_5(CO_3)_2(OH)_6$
Chemical class:	Anhydrous carbonate with hydroxyl or halogen
Chemical type:	$(AB)_5(XO_3)_2Z_q$

Crystal system:	Monoclinic
Mineral group:	Aurichalcite
Space group:	C2/m

Specimen:	BM 1934,975 White chalky massive, with tiny colourless needle-shaped crystals.
Source:	Goodsprings, Clark County, Nevada, U.S.A.
Spectrum ref. no.:	IR2754
Sample medium:	KBr disk
XRD:	1692F = hydrozincite
Composition:	Zn only

Notes

The chalky material gave a much more detailed and well-resolved spectrum than the associated colourless needles (also hydrozincite).

Compare the spectrum with that of hydrozincite ("Dorchester type") IR2755.

References:

1. Jambor J.L. (1966)
    Natural and synthetic hydrozincites.
    Canadian Mineralogist, **8**(5), pp.652-653.

Peak Table cm^{-1}

3300	**692**
3243	**517**
2590	**467**
2395	389
2104	**375**
1767	**316**
1589	280
1548	**265**
1507	
1390	
1364	
1336	
1068	
1048	
953	
893	
836	
738	
710	

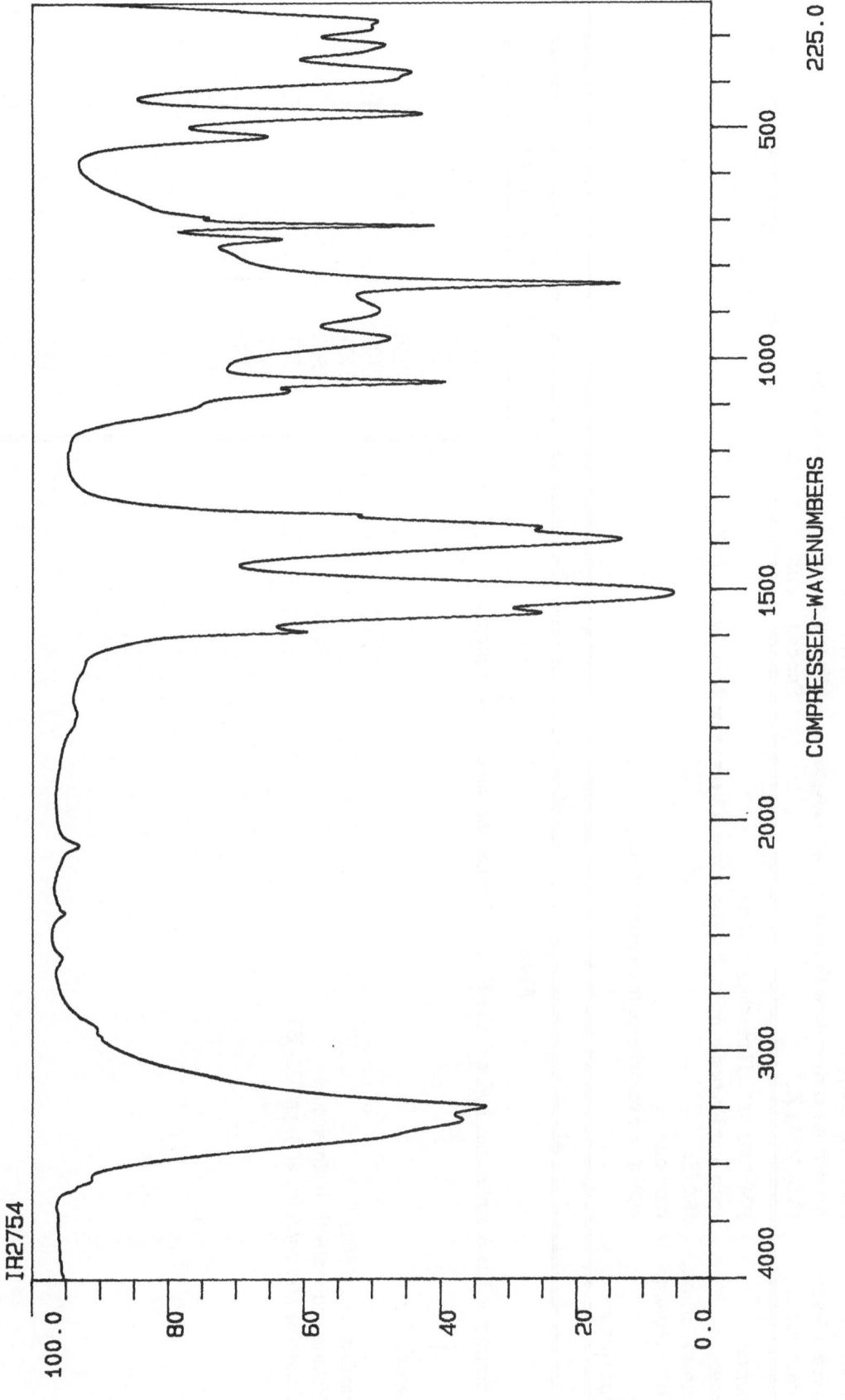

% TRANSMITTANCE

IR2754

100.0

80

60

40

20

0.0

4000   3000   2000   1500   1000   500   225.0

COMPRESSED-WAVENUMBERS

HYDROZINCITE

HYDROZINCITE (dorchester type)

Formula:	$Zn_5(CO_3)_2(OH)_6$
Chemical class:	Anhydrous carbonate with hydroxyl or halogen
Chemical type:	$(AB)_5(XO_3)_2Z_q$
Crystal system:	Monoclinic
Mineral group:	Aurichalcite
Space group:	C2/m

Specimen:	BM 1983,452. Thin white coating.
Source:	Parc and Fucheslas mine, Betws-y-Coed, Caernarvonshire, Wales, U.K.
Spectrum ref. no.:	IR2755
Sample medium:	KBr disk
XRD:	4881F = hydrozincite (dorchester type)
Composition:	

Peak Table cm⁻¹

2359
2584
2361
2336
2102
1774
1508
1387
1047
950
888
834
738
708
474
367
296

Notes

The spectrum is less complex than hydrozincite IR2754, but has the same overall pattern.

References:

1. Jambor J.L. (1966)
    Natural and synthetic hydrozincites.
    Canadian Mineralogist, **8**(5), pp.652-653.

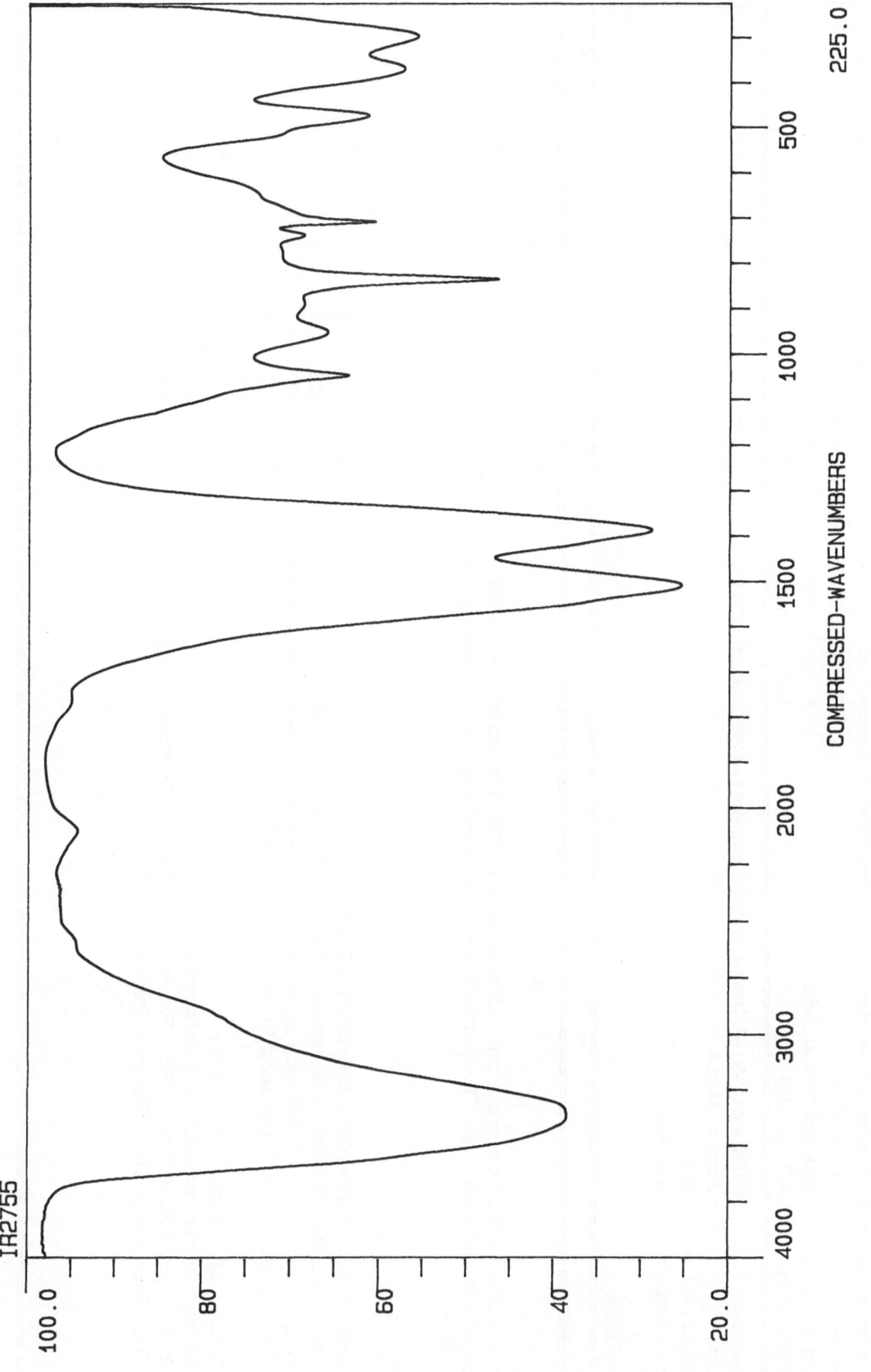

% TRANSMITTANCE

IR2755

COMPRESSED-WAVENUMBERS

HYDROZINCITE

IKAITE ?

Formula:	**CaCO₃·6H₂O**

$Formula:$ **CaCO$_3$·6H$_2$O**

Crystal system: Monoclinic

Chemical class: **Hydrated normal carbonate** **Mineral group:** Nesquehonite

Chemical type: **A(XO$_3$)·xH$_2$O** **Space group:** C2/c or Cc

Specimen: **NHM unregistered. Brown, translucent, bipyramidal isolated crystals.**

Source: **Barrow, Alaska.**

Spectrum ref. no.: **IR2877**

Sample medium: **KBr disk**

XRD:

Composition:

Peak Table cm⁻¹

3467
2515
2387
2258
1797
1641
1421
1082
877
710
680
614
571
316

Notes

Ikaite is unstable at room temperature and is stored below 0°C. This specimen may be partially decomposed to calcite plus water. The spectrum was checked by repeating as a mull which gave an identical result.

References:

1. Kennedy G.L., Hopkins D.M. & Pickthorn W.J. (1987)
Ikaite, the glendonite precursor, in estuarine sediments at Barrow, Arctic Alaska.
In: Dickinson, W.R. (Ed) *Geological Society of America, 1987 annual meeting and exposition. Abstracts with Programs, Geological Society of America,* **19**(7), p.725.

2. Shaikh A.M. & Shearman D.J. (1987)
On ikaite and the morphology of its pseudomorphs.
In: Rodriguez C.R. & Tardy Y.(Eds.) *Geochemistry and mineral formation in the Earth surface. Cons. Super. Invest. Cient.,* Barcelona, Spain, pp.791-803.

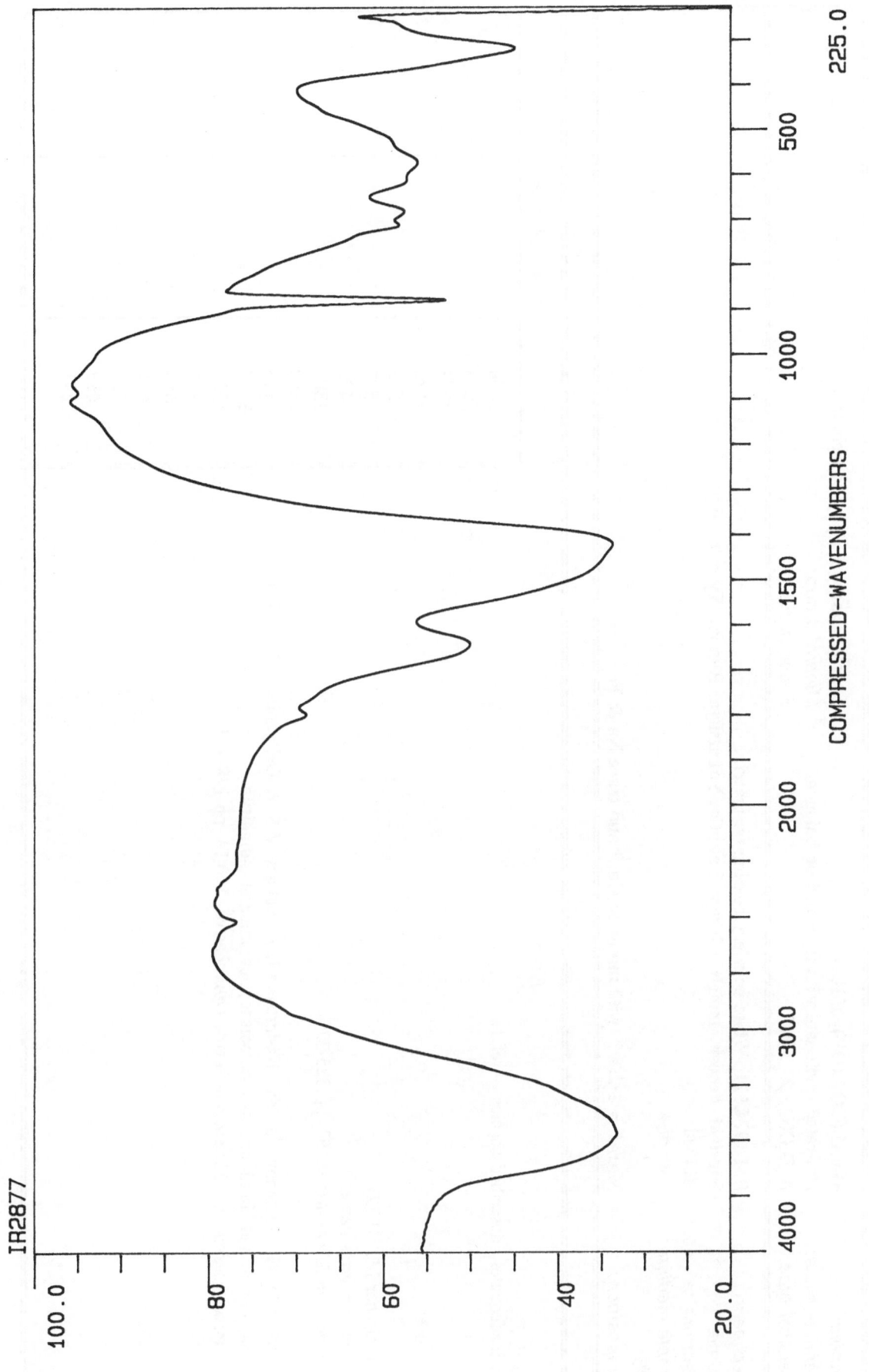

IR2877

% TRANSMITTANCE

100.0

80

60

40

20.0

4000   3000   2000   1500   1000   500   225.0

COMPRESSED-WAVENUMBERS

IKAITE ?

INDIGIRITE ?

Formula:	$Mg_2Al_2(CO_3)_4(OH)_2 \cdot 15H_2O$
Chemical class:	Hydrated carbonate with hydroxyl or halogen
Chemical type:	$A_mB_n(XO_3)_pZ_q \cdot xH_2O$
Crystal system:	Monoclinic ?
Mineral group:	
Space group:	?
Specimen:	BM 1974,521 White powdery coating on matrix.
Source:	Sarylakh deposit, Indigirki River, Yakutia, Yakutskya, Russia. (Type locality).
Spectrum ref. no.:	IR2880
Sample medium:	KBr disk
XRD:	
Composition:	Mg:Al = 1·2:1 ? with minor Si,Ca,P and trace Na & Fe

Peak Table cm⁻¹

3460
2073
1747
1636
1543
1513
1400
1384
1360
1113
1024
975
786
689
621
558
450
243

Notes

"An inadequately described species" - ref.1.

References:

1. Fleischer M. (1972)
    New minerals names.
    American Mineralogist, **57**, pp.325-329.

2. Indolev L.N., Zhdanov Yu. Ya., Kashirtseva K.I., Suknev V.S. & Del'yanidi K.I. (1971)
    Magnesium and aluminum hydrocarbonate; new mineral indighirite.
    Zapiski Vsesoiuznoe Mineralogicheskoe Obshchestvo, **100**(2), pp.178-183.

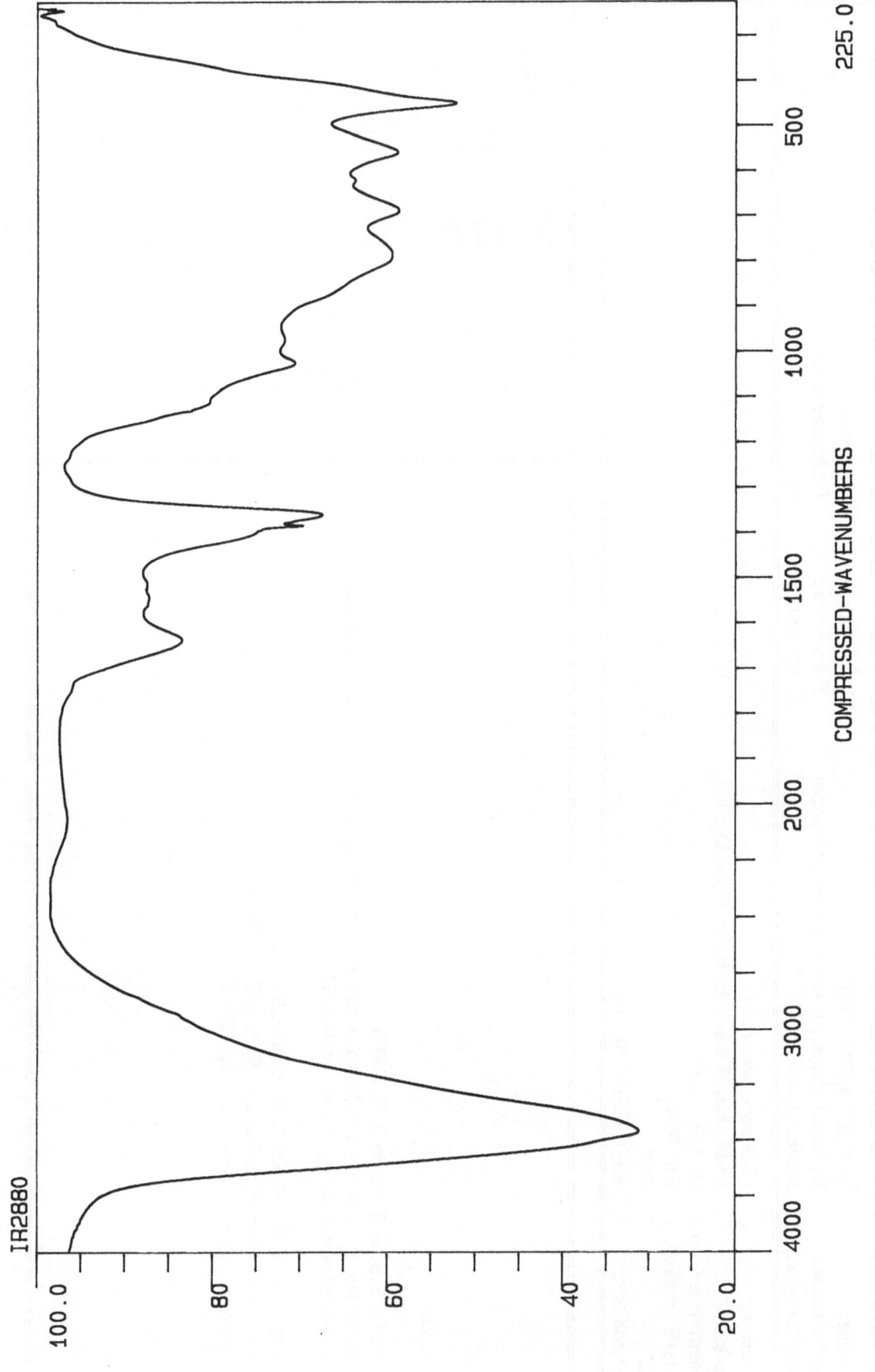

IR2880

% TRANSMITTANCE

100.0

80

60

40

20.0

4000   3000   2000   1500   1000   500   225.0

COMPRESSED-WAVENUMBERS

INDIGIRITE

KAMBALDAITE

Formula:	NaNi$_4$(CO$_3$)$_3$(OH)$_3 \cdot$3H$_2$O
Chemical class:	Hydrated carbonate with hydroxyl or halogen
Chemical type:	Miscellaneous

Crystal system:	Hexagonal
Mineral group:	Hydromagnesite
Space group:	P6$_3$

Specimen:	RMS unregistered.
Source:	Kambalda, Western Australia. (Type locality).
Spectrum ref. no.:	IR2939
Sample medium:	KBr disk
XRD:	4257
Composition:	Major Ni and Na only

Peak Table cm^{-1}

3519	368
3483	340
2927	285
2545	
2499	
2457	
2169	
1806	
1619	
1471	
1395	
1086	
968	
872	
856	
739	
721	
508	
412	

Notes

References:

1. Nickel E.H. & Robinson B.W. (1985)
   Kambaldaite; a new hydrated Ni Na carbonate mineral from Kambalda, Western Australia.
   American Mineralogist, 70(3,4), pp.419-422.

2. Engelhardt L.M., Hall S.R. & White A.H. (1985)
   Crystal structure of kambaldaite, Na$_2$Ni$_8$(CO$_3$)$_6$(OH)$_6 \cdot$6H$_2$O.
   American Mineralogist, 70(3,4), pp.423-427.

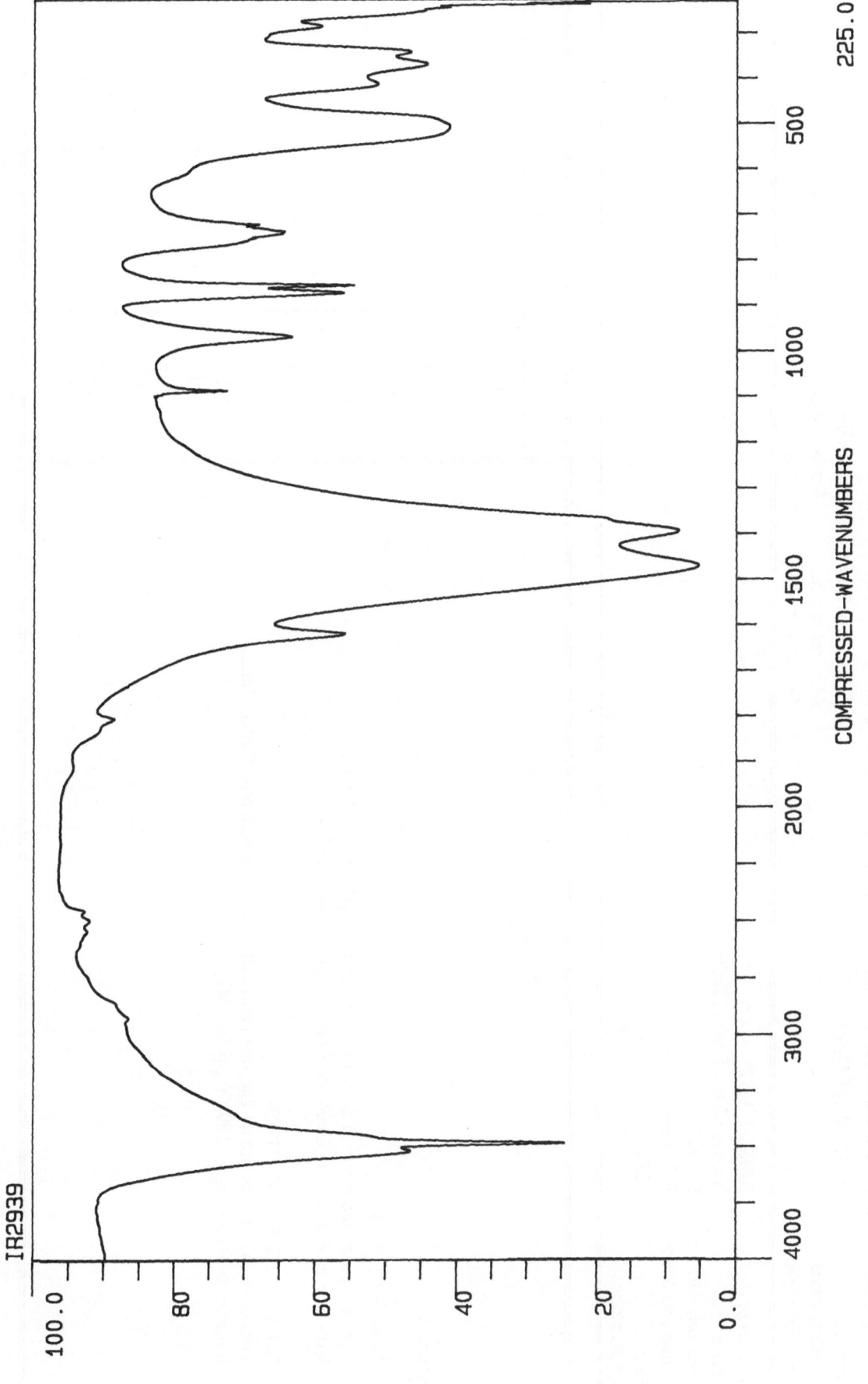

IR2939

% TRANSMITTANCE

COMPRESSED-WAVENUMBERS

KAMBALDAITE

KAMOTOITE-(Y)

Crystal system:	**Monoclinic**
Mineral group:	**Rutherfordine**
Space group:	**P2$_1$/a**

Formula:	**Y$_2$O$_4$(UO$_2$)$_4$(CO$_3$)$_3$·14H$_2$O**
Chemical class:	
Chemical type:	

Specimen:	**RMS RC3692**
Source:	**Kamoto Est., Shaba, Zaïre.**
Spectrum ref. no.:	**IR2921**
Sample medium:	**KBr disk**
XRD:	**4550**
Composition:	

Peak Table cm^{-1}

3423
2929
2857
2457
1865
1731
1606
1537
1364
1195
1151
1122
911
742
660
509
366
280

Notes

References:

1. Cesbron F. (1987)
   New minerals; kamotoite (Y)$_4$UO$_3$(Y,Nd,Gd,Sm,Dy)$_2$O$_3$·3CO$_2$·14·5H$_2$O
   Mineraux et Fossiles, le Guide du Collectionneur, **146**, p.35.

2. Deliens M. & Piret P. (1986)
   Kamotoite (Y), a new uranyl and rare earth carbonate from Kamoto, Shaba, Zaire.
   Bulletin de Minéralogie, **109**(6), pp.643-647.

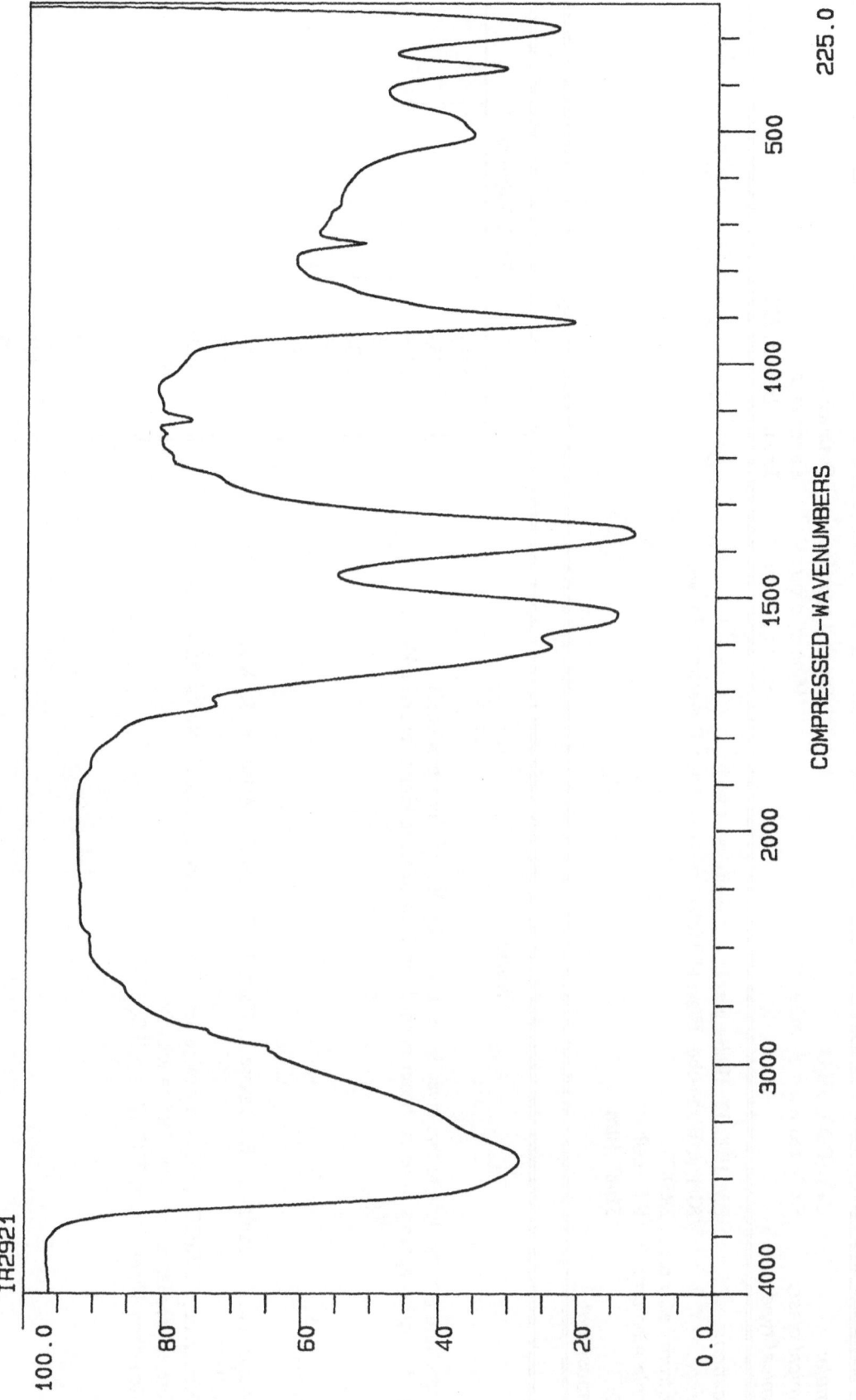

IR2921

% TRANSMITTANCE

100.0   80   60   40   20   0.0

4000   3000   2000   1500   1000   500   225.0

COMPRESSED-WAVENUMBERS

KAMOTOITE- (Y)

KIMURAITE-(Y)

Formula:	CaY$_2$(CO$_3$)$_4$·6H$_2$O
Chemical class:	Hydrated normal carbonate
Chemical type:	
Crystal system:	Orthorhombic
Mineral group:	Lanthanite
Space group:	Imm2, Immm, I222

Specimen:	BM 1986,32 White pearly aggregate on matrix.
Source:	Kirigo, Hizen-cho, Higashi Matsuura-Gun, Saga Prefecture, Japan.
Spectrum ref. no.:	IR3071
Sample medium:	KBr disk
XRD:	5168F (std)
Composition:	

Peak Table cm^{-1}

3431
2963
2925
2856
2382
2296
1777
1636
1522
1392
1085
1061
855
842
744
688
545
295

Notes

The spectrum is very close to that given in ref.1 for **lokkaite** (poorly reproduced).
X-ray diffraction appears to be the better method for distinguishing kimuraite and lokkaite.

References:

1. Nagashima K., Miyawaki R., Takase J., Nakai I., Sakurai K., Matsubara S., Kato A. & Iwano S. (1986)
Kimuraite, CaY$_2$(CO$_3$)$_4$·6H$_2$O, a new mineral from fissures in an alkali olivine basalt from Saga Prefecture, Japan, and new data on lokkaite.
American Mineralogist, **71**(7,12), pp.1028-1033.

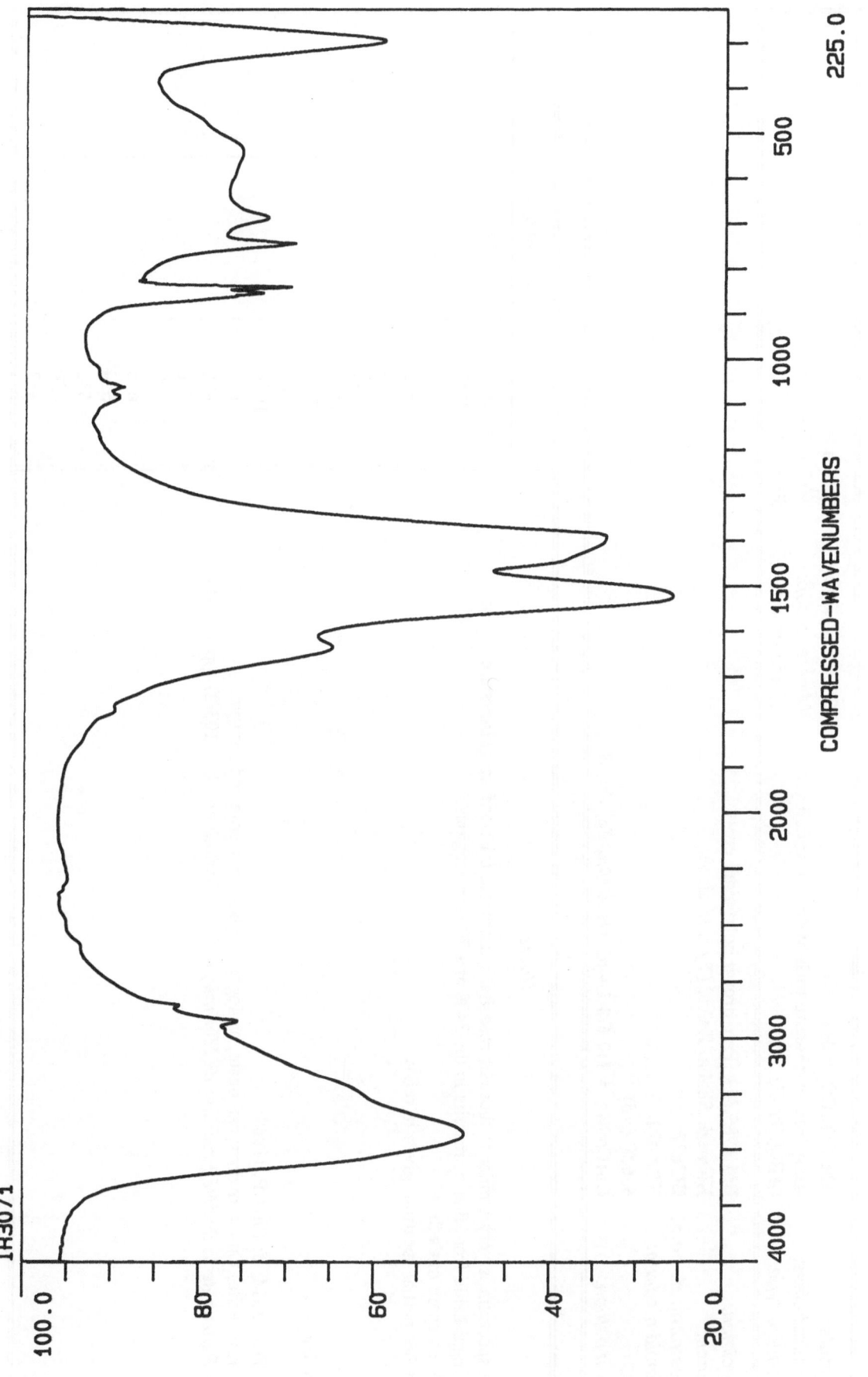

IR3071

% TRANSMITTANCE

COMPRESSED-WAVENUMBERS

KIMURAITE- (Y)

KOLWEZITE

Formula:	$(Cu,Co)_2(CO_3)(OH)_2$
Chemical class:	Anhydrous carbonate with hydroxyl or halogen
Chemical type:	$(AB)_2(XO_3)Z_q$
Crystal system:	Triclinic ?
Mineral group:	Malachite
Space group:	P1 or P$\bar{1}$

Specimen:	BM 1985,311 Pale brown botryoidal crust.
Source:	Kolwezi, Shaba, Zaïre. (Type locality).
Spectrum ref. no.:	IR2827
Sample medium:	KBr disk
XRD:	6562F (std)
Composition:	Cu:Co:Mg = 1:0·6:0·1 with trace Mn, Fe, Zn, S

Notes

The spectrum is very similar to those of rosasite, glaukosphaerite and mcguinnessite, but is distinguishable from that of rosasite in the 3400 and 700 cm^{-1} regions.
Also compare **malachite.**
Matches partial spectrum given in ref.1.

References:

1. Deliens M. & Piret P. (1980)
    Kolwezite, Cu Co hydroxycarbonate, analogue of glaucosphaerite and rosasite.
    Bulletin de la Société Française de Minéralogie et de Cristallographie, **103**(2), pp.179-184.

Peak Table cm^{-1}

3474	277
3249	
2561	
2392	
1774	
1543	
1419	
1371	
1103	
1050	
854	
828	
740	
709	
676	
554	
532	
420	
333	

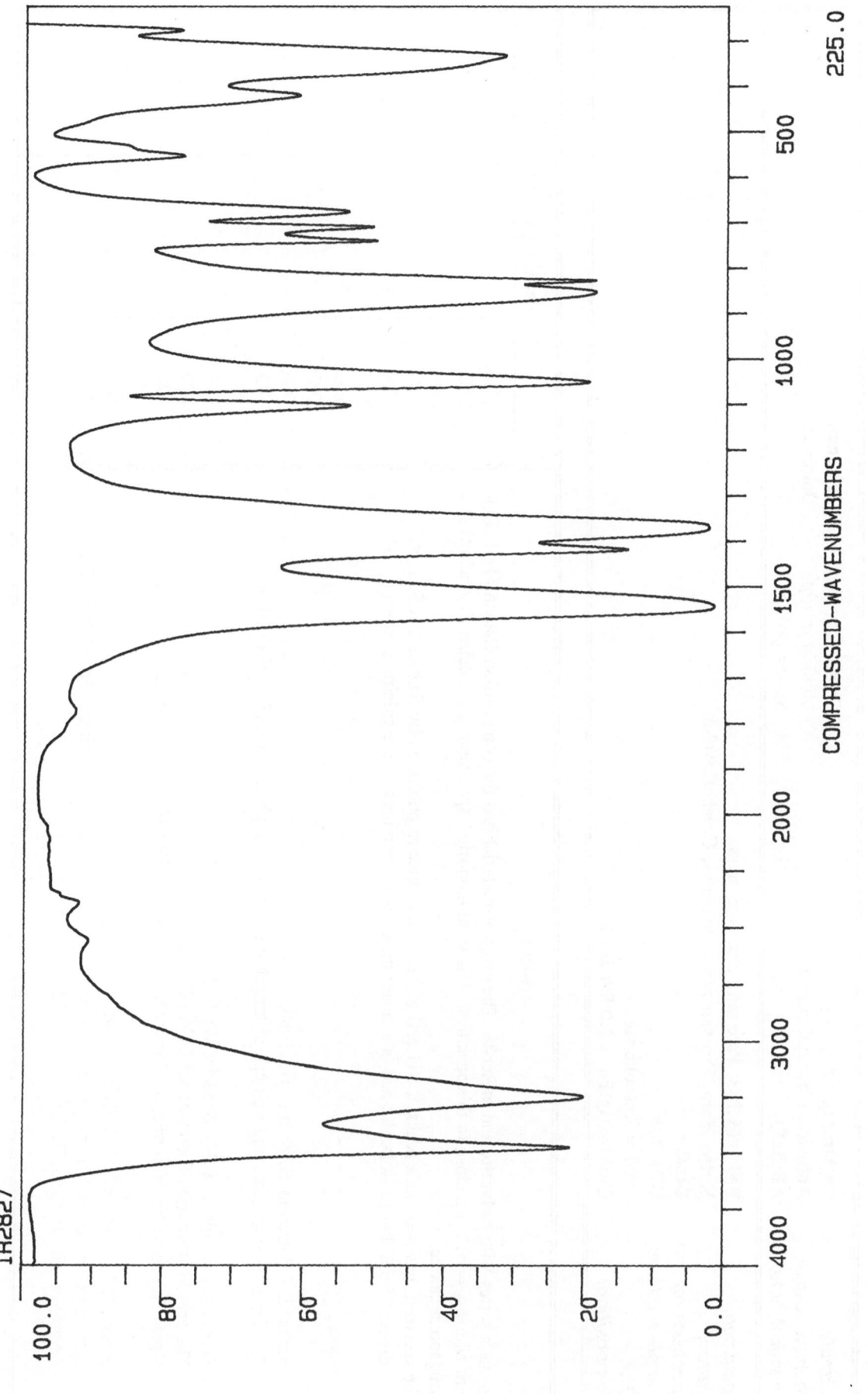

IR2827

% TRANSMITTANCE

4000　　　3000　　　2000　　　1500　　　1000　　　500　　225.0

COMPRESSED-WAVENUMBERS

KOLWEZITE

KUTNOHORITE

Formula:	Ca(Mn,Mg,Fe)(CO₃)₂	
Chemical class:	Anhydrous normal carbonate	
Chemical type:	AB(XO₃)₂	

Crystal system:	Trigonal
Mineral group:	Dolomite
Space group:	R$\bar{3}$

Specimen:	BM 1969,283 Pale pink cleavage mass.
Source:	Kutna Hora, Prazska zupa, Bohemia, Czechoslovakia.
Spectrum ref. no.:	IR2674
Sample medium:	KBr disk
XRD:	1500 = kutnohorite
Composition:	Ca:Mn:Mg:Fe = 1:0·8:0·2:0·1

Notes

Forms a series with **dolomite** and **ankerite**. The original material had the composition Ca(Mn,Mg)(CO₃)₂ with Mn:Mg = 5:2 i.e close to this specimen. Many 'kutnohorite' specimens are calcian rhodochrosite or manganoan calcite.

The spectrum has two extra peaks at 841 and 679 cm⁻¹ c.f. spectra published by Sadtler and Suhner. It is distinct from that of **dolomite** and is a closer match with members of the **calcite** group.

References:

1. Farkas L., Bolzenius B. & Will G. (1988)
    Powder diffraction data and unit cell of kutnahorite. *Powder Diffraction,* **3**(3), pp.172-174.

2. Farkas L., Bolzenius B.H., Schaefer W. & Will G. (1988)
    The crystal structure of kutnahorite CaMn(CO₃)₂.
    Neues Jahrbuch für Mineralogie, Monatshefte, (12), pp.539-546.

3. Peacor D.R., Essene E.J. & Gaines A.M. (1987)
    Petrologic and crystal chemical implications of cation order disorder in kutnahorite [CaMn(CO₃)₂].
    American Mineralogist, **72**(3,4), pp.319-328.

Peak Table cm⁻¹

[3530]
2982
2864
2601
2504
1804
1424
1089
1052
873
841
721
679
522
474
346
321

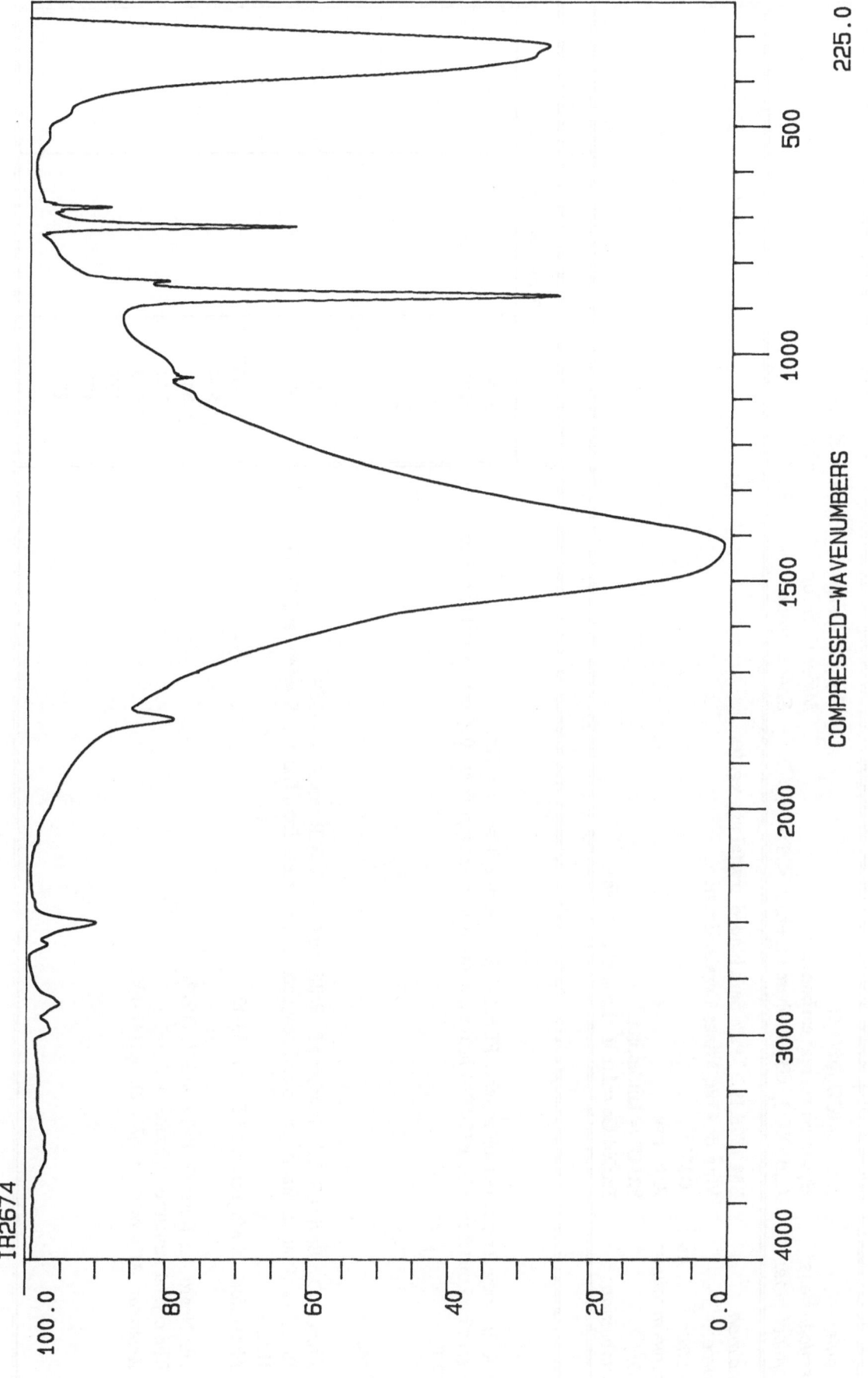

% TRANSMITTANCE

COMPRESSED-WAVENUMBERS

IR2674

KUTNOHORITE

LANTHANITE-(La)

Formula:	$(La,Ce)_2(CO_3)_3 \cdot 8H_2O$
Chemical class:	Hydrated normal carbonate
Chemical type:	$A_mB_n(XO_3)_p \cdot xH_2O$ where $(m+n):p < 1:1$

Crystal system:	Orthorhombic
Mineral group:	Lanthanite
Space group:	Pbnb

Specimen:	BM 1975,192 Pale pink bladed crystal aggregates.
Source:	Vina de Uba, Minas Gerais, Brazil.
Spectrum ref. no.:	IR2893
Sample medium:	KBr disk
XRD:	8234F = lanthanite
Composition:	La:Nd:Ce ≈ 1:1:0 (La slightly > Nd)

Peak Table cm⁻¹

3359
3205
2472
2414
2274
1849
1761
1484
1377
1079
[1036]?
874
849
748
679
657
472
287

NOTES

The X-Ray powder photograph matches PDF30-678 lanthanite (also La ≈ Nd).
The spectrum matches Suhner (5-59A) lanthanite, but has an extra peak at 1036 cm⁻¹ possibly due to impurity.

References:

1. Atencio D., Bevins R.E., Fleischer M., Williams C.T. & Williams P.A. (1989)
    Revision of the lanthanite group and new data for specimens from Bastnäs, Sweden, and Bethlehem U.S.A.
    Mineralogical Magazine, **53**(5), pp.639-42

2. Dal Negro A., Rossi G. & Tazzoli V. (1977)
    The crystal structure of lanthanite.
    American Mineralogist, **62**(1,2), pp.142-146.

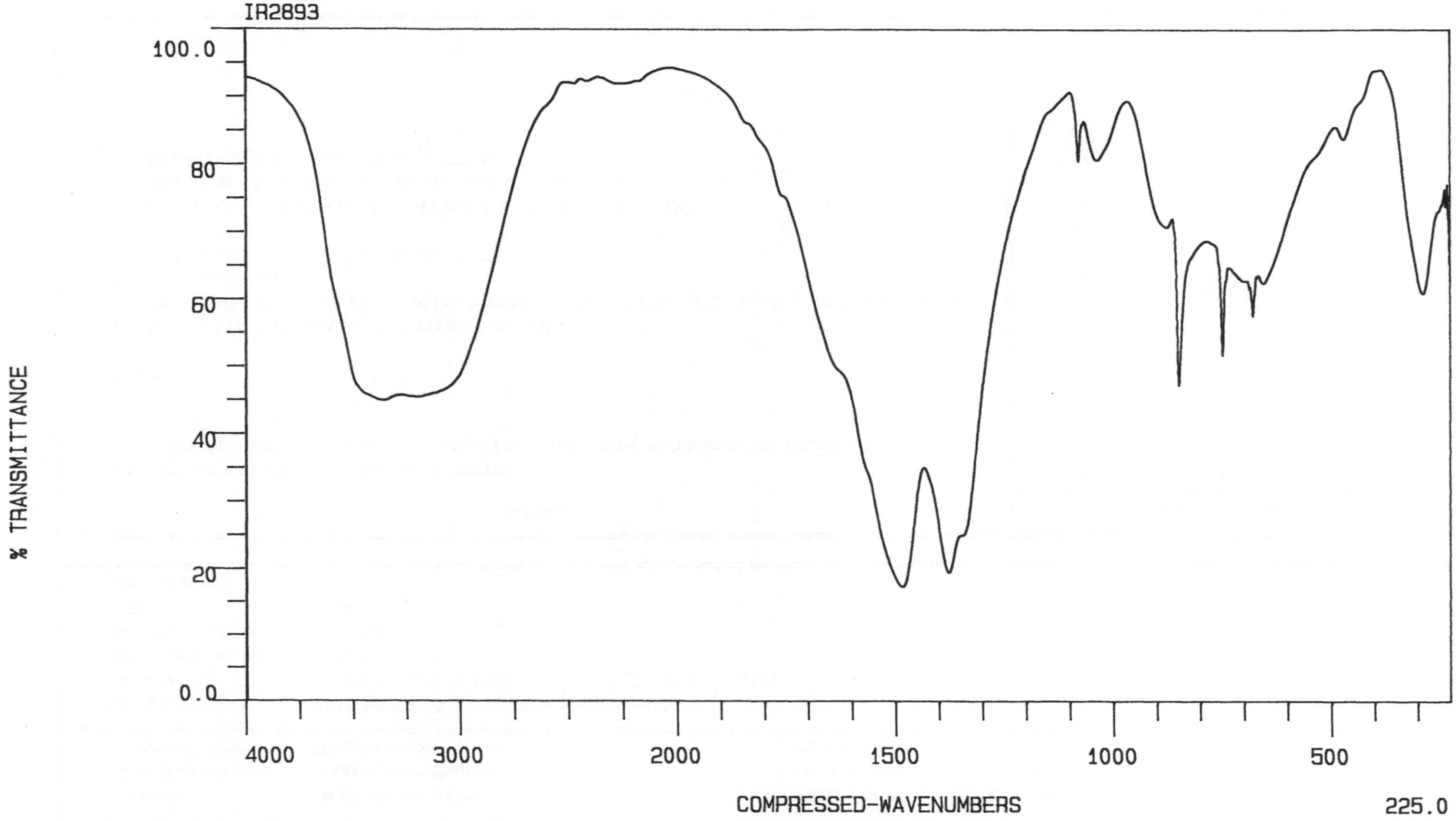

IR2893

% TRANSMITTANCE

100.0   80   60   40   20   0.0

4000   3000   2000   1500   1000   500   225.0

COMPRESSED-WAVENUMBERS

LANTHANITE- (La)

LEADHILLITE

Crystal system:	**Monoclinic**
Mineral group:	**Susannite**
Space group:	**P2₁/a**

Formula:	**Pb₄(SO₄)(CO₃)₂(OH)₂**
Chemical class:	**Compound carbonate**
Chemical type:	**Miscellaneous**
Specimen:	**RMS 1908.13.5 Pale yellow platy crystals.**
Source:	**Leadhills, Lanarkshire, Scotland, U.K.** (Type locality).
Spectrum ref. no.:	**IR2929**
Sample medium:	**KBr disk**
XRD:	**3909**
Composition:	

Peak Table cm⁻¹

3472	394
3380	373
2926	
2857	
2413	
1735	
1629	
1401	
1088	
1055?	
1042	
964	
859	
840	
706	
681	
632	
601	
423	

NOTES

Trimorphous with susannite and macphersonite.

The spectrum is very similar to, but distinguishable from, those of susannite and macphersonite.

References:

1. Russell J.D., Fraser A.R. & Livingstone A. (1984)
The infrared absorption spectra of the three polymorphs of PbSO₄(CO₃)₂(OH)₂ (leadhillite, susannite, and macphersonite).
Mineralogical Magazine. **48**(2), pp.295-7.

2. Russell J.D., Milodowski A.E., Fraser A.R. & Clark D.R. (1983)
New IR and XRD data for leadhillite of ideal composition.
Mineralogical Magazine, **47**(3), pp.371-5.

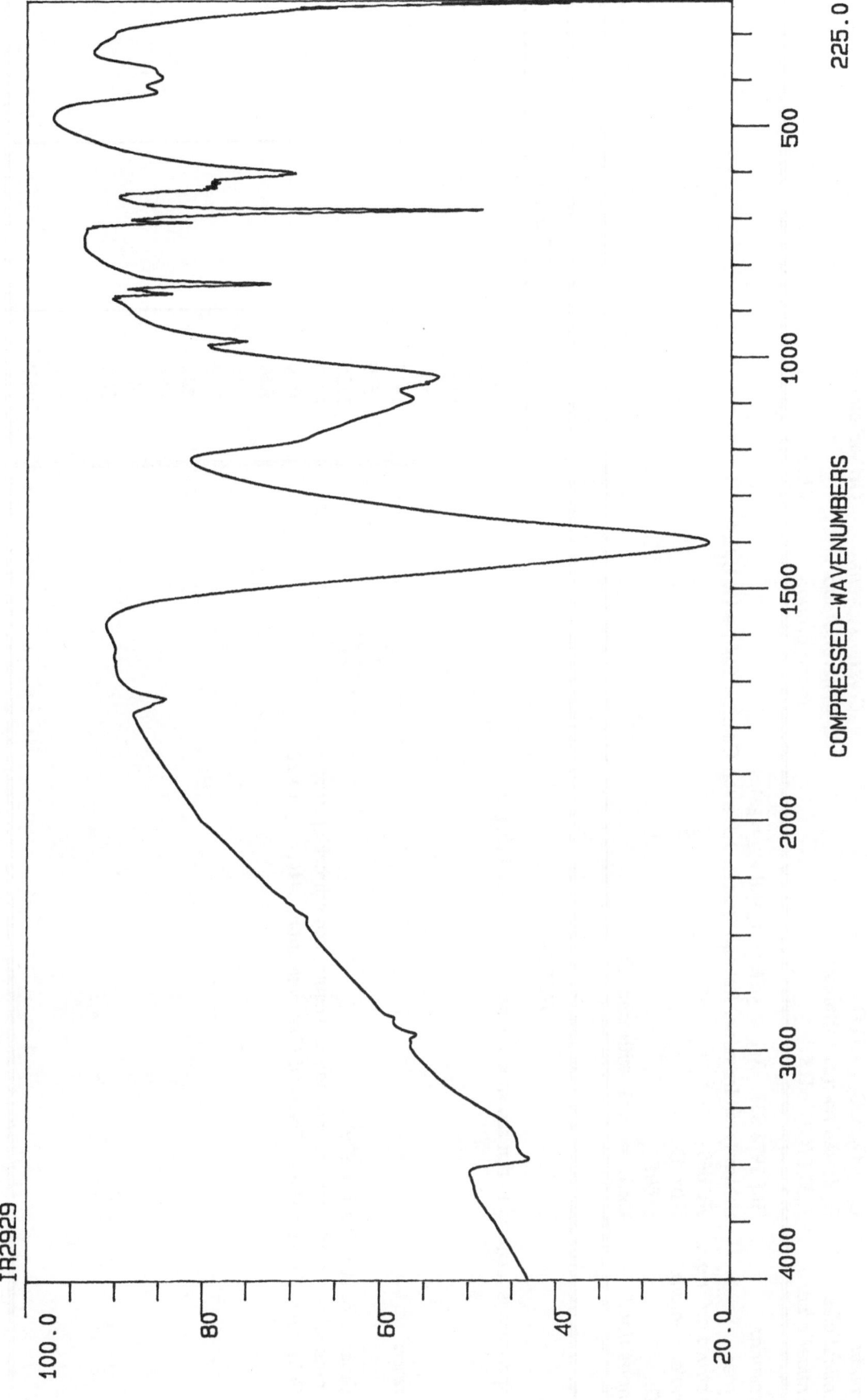

IR2929

% TRANSMITTANCE

COMPRESSED-WAVENUMBERS

LEADHILLITE

LIEBIGITE

Crystal system:	Orthorhombic
Mineral group:	
Space group:	Bbam

Formula:	$Ca_2(UO_2)(CO_3)_3 \cdot 11H_2O$
Chemical class:	Hydrated normal carbonates
Chemical type:	$A_m B_n (XO3)_p \cdot xH_2O$
Specimen:	BM 1978,338 Bright yellow crystal aggregates
Source:	Schwartzwalder mine, Ralston Buttes, Jefferson County, Colorado, U.S.A.
Spectrum ref. no.:	IR2860
Sample medium:	KBr disk
XRD:	8100F
Composition:	Ca:U = 1·6:1 with trace Si

Peak Table cm⁻¹

3484		
2622		
1624		
1549		
1515		
1379		
1155		
1070		
893		
846		
823		
799		
742		
523		
316		
286		
242		
238?		
233?		

NOTES

The spectrum is identical to that shown in Suhner (5-36 A) for liebigite.

References:

1. Urbanec Z. & Čejka J. (1979)
    Infrared spectra of liebigite, andersonite, voglite, and schröckingerite.
    Collection of Czechoslovak Chemical Communications, **44**(1), pp.10-23.

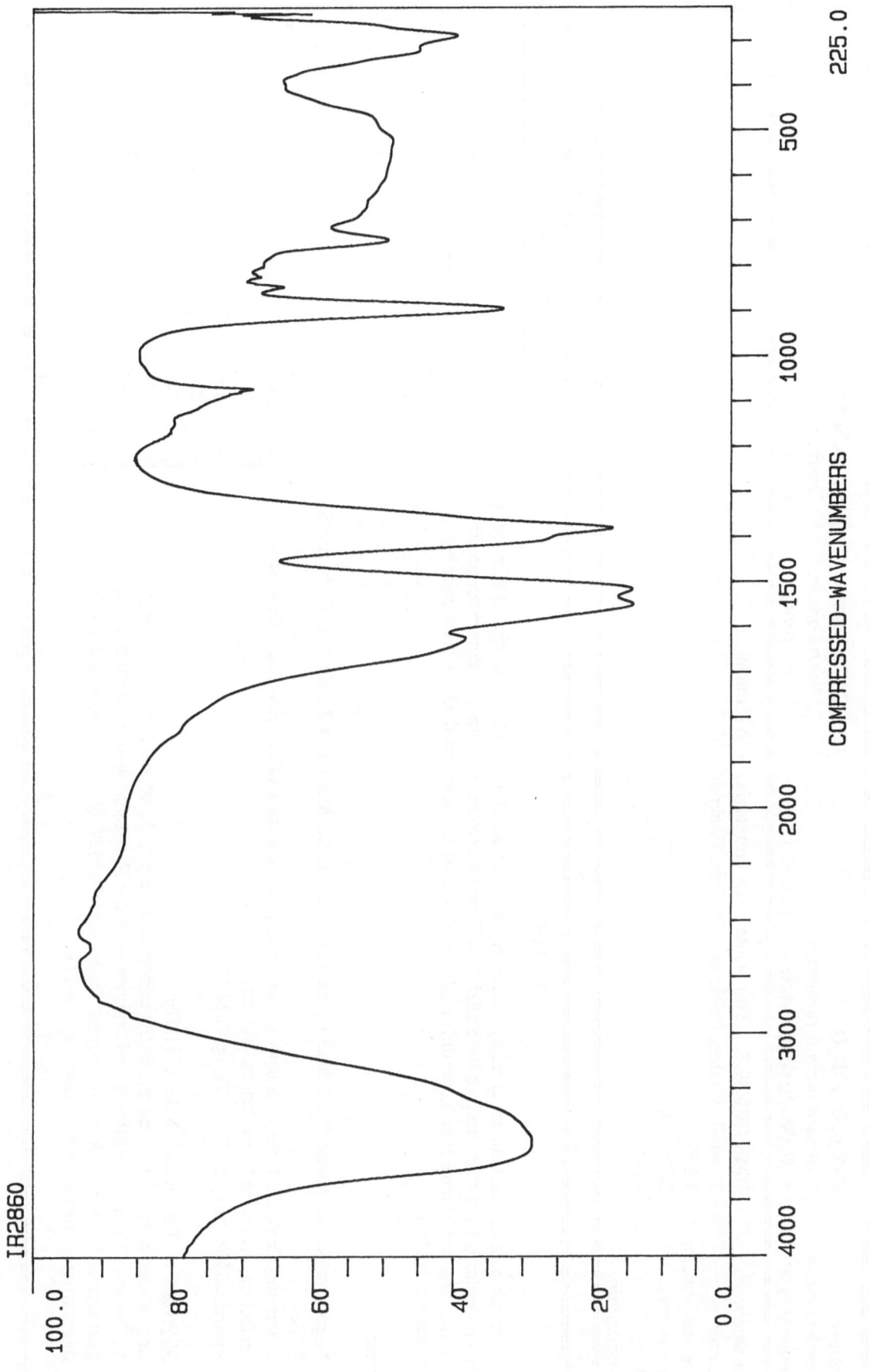

IR2860

% TRANSMITTANCE

COMPRESSED-WAVENUMBERS

LIEBIGITE

LOKKAITE

Formula:	$CaY_4(CO_3)_7 \cdot 9H_2O$
Chemical class:	Hydrated normal carbonate
Chemical type:	$A_m B_n (XO_3)_p \cdot xH_2O$ where $(m+n):p < 1:1$

Crystal system:	Orthorhombic
Mineral group:	Lanthanite
Space group:	Cmmm

Specimen:	RMS 1980.49.5. Tiny white spots on matrix, with kainosite.
Source:	Evans-Lou mine, Poltimore, Quebec, Canada.
Spectrum ref. no.:	IR2951
Sample medium:	KBr disk
XRD:	4141
Composition:	

Peak Table cm⁻¹

3403
2592
1865
1824
1784
1636
1510
1410
1091
1066
865
850
837
765
721
687
583
464
301

NOTES

The spectrum matches that shown in ref.1. (poorly reproduced) which is very close to that of **kimuraite**. Another specimen from this locality gave a slightly different spectrum, also close to kimuraite, possibly due to the presence of tengerite. X-ray diffraction appears to be the better method for distinguishing kimuraite and lokkaite.

References:

1. Nagashima K., Miyawaki R., Takase J., Nakai I., Sakurai K., Matsubara S., Kato A. & Iwano S. (1986)
    Kimuraite, $CaY_2(CO_3)_4 \cdot 6H_2O$, a new mineral from fissures in an alkali olivine basalt from Saga Prefecture, Japan, and new data on lokkaite.
    American Mineralogist, 71(7,12), pp.1028-1033.

2. Miyawaki R, Takase J. & Nakai I. (1986)
    Crystal chemistry of hydrous rare earths carbonate minerals; the crystal structure of tengerite.
    In: Prewitt C.T. (Ed.) *Abstracts and programme of the Fourteenth general meeting of the International Mineralogical Association. Papers and Proceedings of the General Meeting International Mineralogical Association*, p.173.

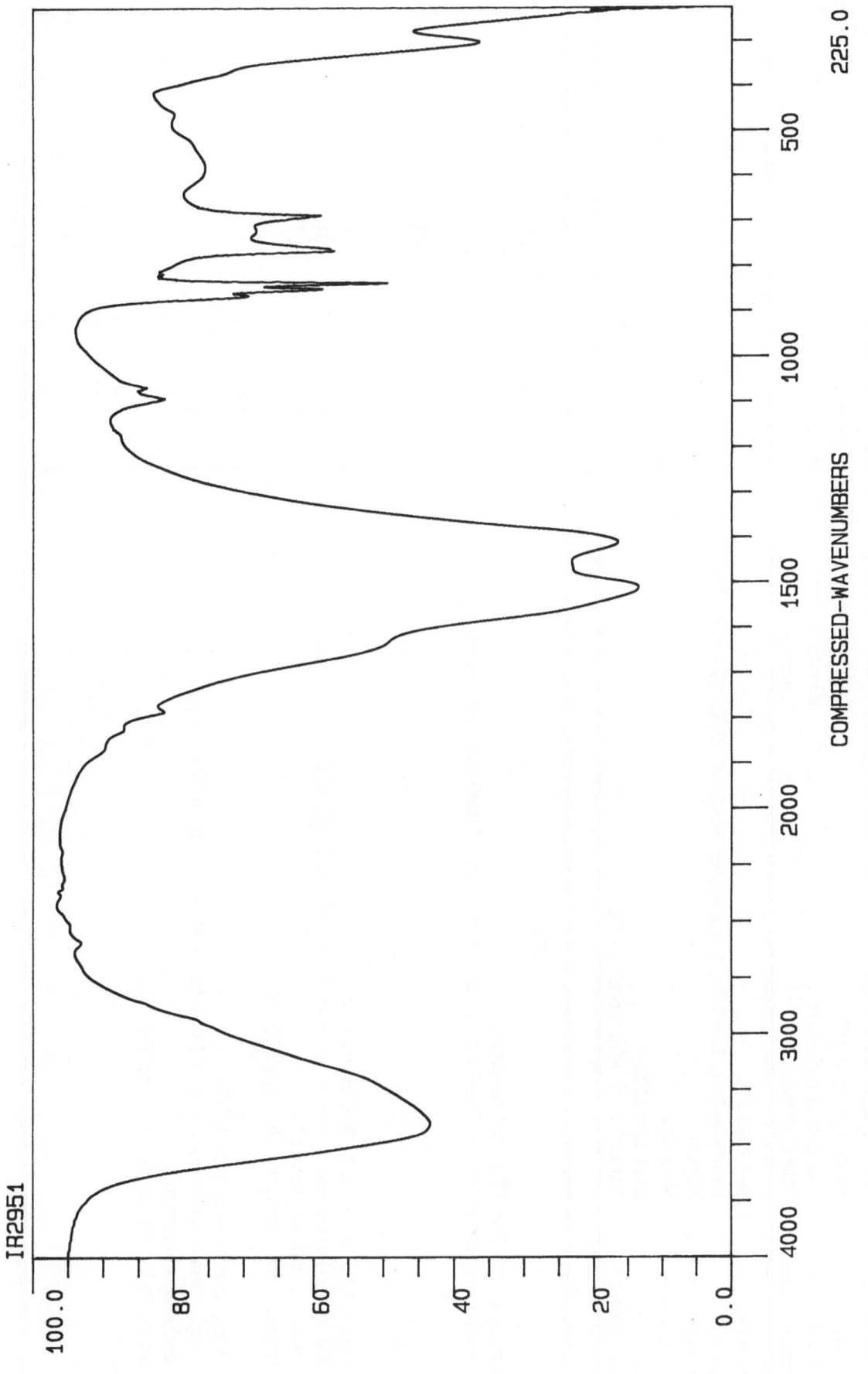

IR2951

% TRANSMITTANCE

100.0
80
60
40
20
0.0

4000
3000
2000
1500
1000
500
225.0

COMPRESSED-WAVENUMBERS

LOKKAITE—(Y)

MACPHERSONITE

Crystal system:	Orthorhombic
Mineral group:	Susannite
Space group:	$P2_12_12_1$

Formula:	$Pb_4(SO_4)(CO_3)_2(OH)_2$
Chemical class:	Compound carbonate
Chemical type:	Miscellaneous
Specimen:	RMS 721.34.
Source:	Leadhills Dod, Leadhills, Lanarkshire, Scotland, U.K. (Type locality).
Spectrum ref. no.:	IR2928
Sample medium:	KBr disk
XRD:	2191 and 2192
Composition:	Major Pb, S with trace Cu, Cd

Peak Table cm^{-1}

3482	701
3432	691
2926	687
2856	681
2416	626
1755	616
1734	588
1632	388
1410	325
1362	
1147	
1135?	
1062	
967	
857	
841	
796	
717	
708	

NOTES

Trimorphous with **leadhillite** and **susannite**.

The spectrum is very similar to, but distinguishable from, those of **leadhillite** and **susannite**.

References:

1. Russell J.D., Fraser A.R. & Livingstone A. (1984)
   The infrared absorption spectra of the three polymorphs of $PbSO_4 (CO_3)_2 (OH)_2$ (leadhillite, susannite, and macphersonite).
   Mineralogical Magazine, **48**(2), pp.295-297.

2. Livingstone A. & Sarp H. (1984)
   Macphersonite, a new mineral from Leadhills, Scotland, and Saint Prix, France; a polymorph of leadhillite and susannite.
   Mineralogical Magazine, **48**(2), pp.277-282.

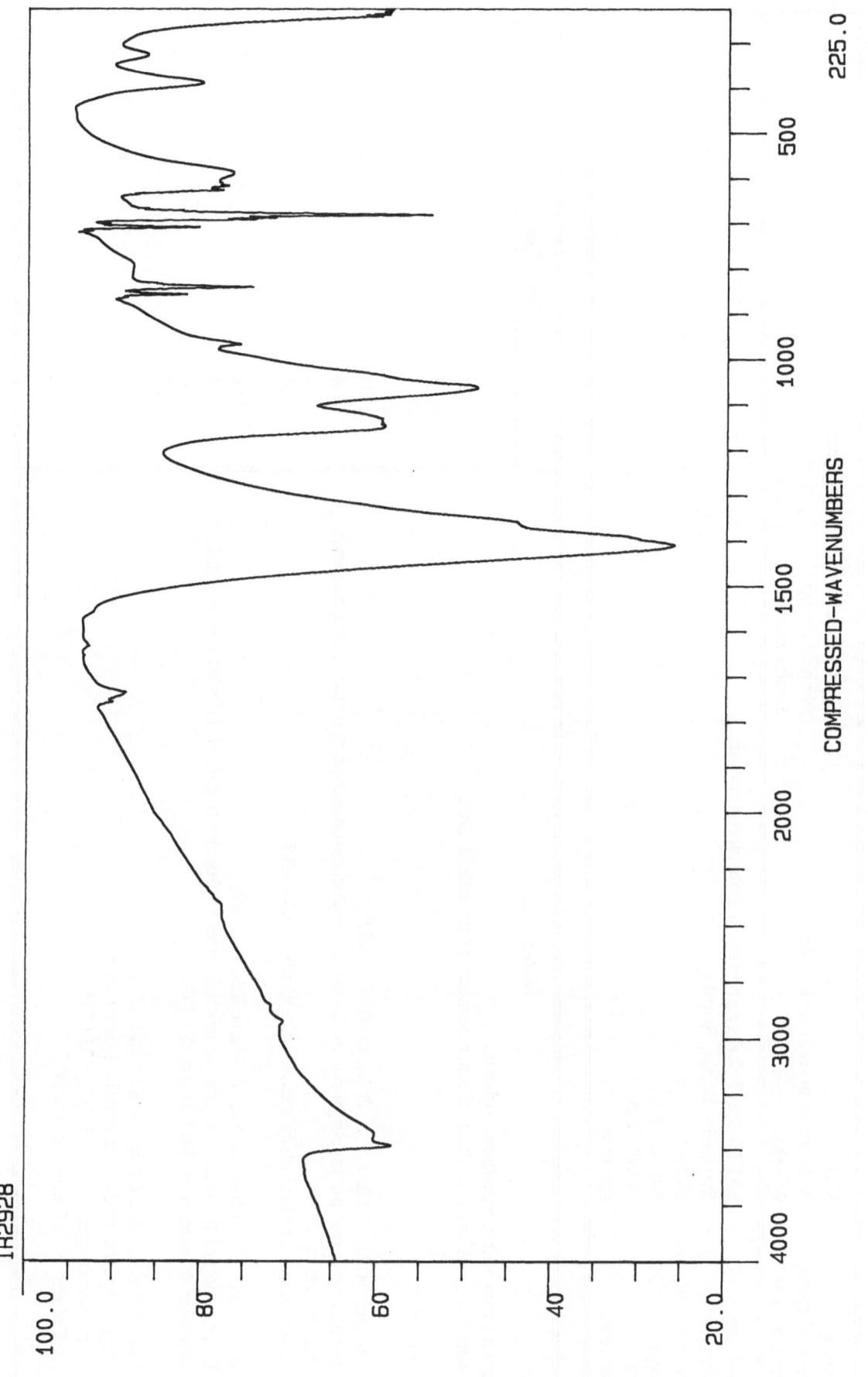

IR2928

% TRANSMITTANCE

100.0   80   60   40   20.0

4000   3000   2000   1500   1000   500   225.0

COMPRESSED-WAVENUMBERS

MACPHERSONITE

MAGNESITE

Formula:	MgCO$_3$
Chemical class:	Anhydrous normal carbonate
Chemical type:	A(XO$_3$)
Crystal system:	Trigonal
Mineral group:	Calcite
Space group:	R$\bar{3}$c

Specimen:	BM 1948,547 Large grey/white translucent rhomb.
Source:	Brumado, Bahia, Brazil.
Spectrum ref. no.:	IR2648
Sample medium:	KBr disk
XRD:	7383F (std)
Chemistry:	Mg only

Peak Table cm^{-1}

[3425]
3051
2922
2619
2537
1831
1446
1093
886
856
748
384
306
258

Notes

Forms a series with **gaspeite** and **siderite**.
Compare the spectrum with those of other members of the **calcite** group.

References:

1. Bottcher M.E., Gehlken P.L. & Usdowski E. (1992)
    Infrared spectroscopic investigations of the calcite-rhodochrosite and parts of the calcite-magnesite mineral series.
    Contributions to Mineralogy and Petrology, **109**, pp.304-306.

2. Dubrawski J.V., Channon A.L. & Warne S.St.J. (1989)
    Examination of the siderite-magnesite mineral series by Fourier transform infrared spectroscopy.
    American Mineralogist, **74**(1,2), pp.187-190.

3. Peng Wenshi., Liu Gaokui & Ke Liqin. (1985)
    Infrared spectra study of magnesite siderite series.
    Acta Mineralogica Sinica, **5**(3), pp.229-233.
    (In Chinese with English summary).

IR2648

% TRANSMITTANCE

COMPRESSED-WAVENUMBERS

MAGNESITE

MALACHITE

Formula:	$Cu_2(CO_3)(OH)_2$
Chemical class:	Anhydrous carbonate with hydroxyl or halogen
Chemical type:	$(AB)_2(XO_3)Z_q$

Crystal system:	Monoclinic
Mineral group:	Malachite (rosasite)
Space group:	P2/a

Specimen:	BM 28043 Silky green needles on limonite.
Source:	Olonetz, Siberia, Russia
Spectrum ref. no.:	IR2734
Sample medium:	KBr disk.
XRD:	4176F (std)
Composition:	Cu only

Peak Table cm⁻¹

3404	**525**
3313	507
2925	**429**
2539	355
2423	**327**
2075	**301**
1841	
1804	
1494	
1421	
1390	
1097	
1047	
875	
822	
778	
749	
713	
571	

NOTES

The spectrum is similar to those of other members of the rosaite group i.e. **glaukosphaerite, kolwezite, mcguinnessite & rosasite.**
Peak assignments are given in ref.2.

References:

1. Timokhina L.V., Balitskii V.S., Shaposhnikov A.A., Bublikova T.M., Kovalenko V.S., Akhmetova G.L., Dubovskii A.B., Andreeva T.G. & Shironina T.V. (1983)
   Physicochemical investigations of synthetic malachite.
   Soviet Physics, Doklady, **28**, pp.429-30.

2. Goldsmith J.A. & Ross S. (1968)
   The infra red spectra of azurite and malachite.
   Spectrochimica Acta, **24**(A), pp.2131-7

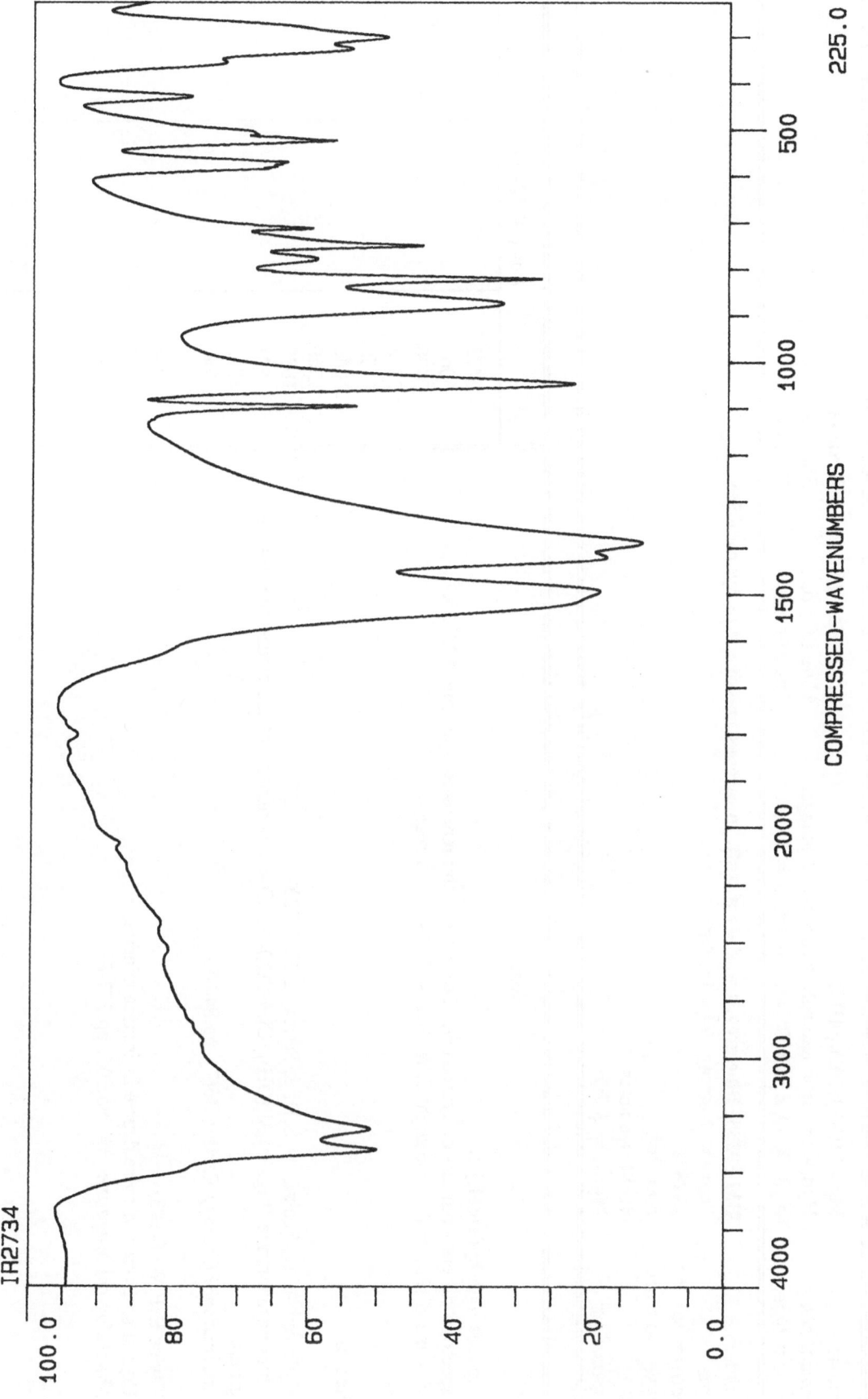

IR2734

% TRANSMITTANCE

COMPRESSED-WAVENUMBERS

MALACHITE

225.0

MANASSEITE

Formula:	$Mg_6Al_2(CO_3)(OH)_{16}\cdot4H_2O$
Chemical class:	Hydrated carbonate with hydroxyl or halogen
Chemical type:	$A_mB_n(XO_3)_pZ_q\cdot xH_2O$ with (m+n):p = 8 : 1

Crystal system:	Hexagonal
Mineral group:	Sjögrenite
Space group:	$P6_3/mmc$

Specimen:	BM 89358 Blue/grey, translucent, soft, micaceous massive with hydrotalcite.
Source:	Snarum, Norway. (type locality)
Spectrum ref. no.:	IR2852
Sample medium:	KBr disk
XRD:	8068F see notes.
Composition:	Mg:Al = 3·7:2

Peak Table cm^{-1}

3473
3067
2438
2336
1752
1640
1366
1080
928
863
768
675
555
445
393

Notes

Dimorphous with **hydrotalcite**.
The spectrum differs from that of hydrotalcite. See notes with **manasseite** spectrum IR2793.
X-ray = manasseite with some slight difference in 1·50-1·55 Å region.

References:

1. Kashayev A.A., Feoktistov G.D. & Petrova S.V. (1983)
   Chlormagaluminite $(Mg,Fe)_4Al_2(OH)_{12}(Cl,\frac{1}{2}CO_3)_2\cdot2H_2O$ a new mineral of the manasseite sjögrenite group.
   International Geology Review, **25**(7), pp.848-53.

2. Taylor H.F.W. (1973)
   Crystal structures of some double hydroxide minerals.
   Mineralogical Magazine. **39**, No.304, pp.377-89.

IR2852

% TRANSMITTANCE

100.0 80 60 40 20 0.0

COMPRESSED-WAVENUMBERS

4000 3000 2000 1500 1000 500 225.0

MANASSEITE

MANASSEITE

Formula:	Mg$_6$Al$_2$(CO$_3$)(OH)$_{16}$·4H$_2$O
Chemical class:	Hydrated carbonate with hydroxyl or halogen
Chemical type:	A$_m$B$_n$(XO$_3$)$_p$Z$_q$·xH$_2$O with (m+n):p = 8:1

Crystal system:	**Hexagonal**
Mineral group:	**Sjögrenite**
Space group:	**P6$_3$/mmc**

Specimen:	**BM 1982,446 Orange bipyramidal crystals in calcite, with magnetite.**
Source:	**Jacupiranga Apatite Quarry, São Paulo, Brazil.**
Spectrum ref. no.:	**IR2793**
Sample medium:	**KBr disk**
XRD:	**2886F = manasseite**
Composition:	**Mg:Al ≈ 3·3:2 ? with trace Fe**

Peak Table cm^{-1}

3439
2926
2859
2412
1536
1400
1355
1081
945
786
675
552
451
394
245?

Notes

Dimorphous with **hydrotalcite**. Distinguishable from other members of the sjögrenite group. See also **manasseite** from Snarum, Norway, IR2852 - the two manasseite spectra are almost identical despite their very different physical forms and association. The Jacupiranga manasseite is found in an alkaline igneous environment and may be related to chlormagaluminite.

References:

1. Kashayev A.A., Feoktistov G.D. & Petrova S.V. (1983)
    Chlormagaluminite (Mg,Fe)$_4$Al$_2$(OH)$_{12}$ (Cl,½CO$_3$)$_2$·2H$_2$O a new mineral of the manasseite sjögrenite group.
    International Geology Review, **25**(7), pp.848-53.

2. Taylor H.F.W. (1973)
    Crystal structures of some double hydroxide minerals.
    Mineralogical Magazine, **39**(304), pp.377-89.

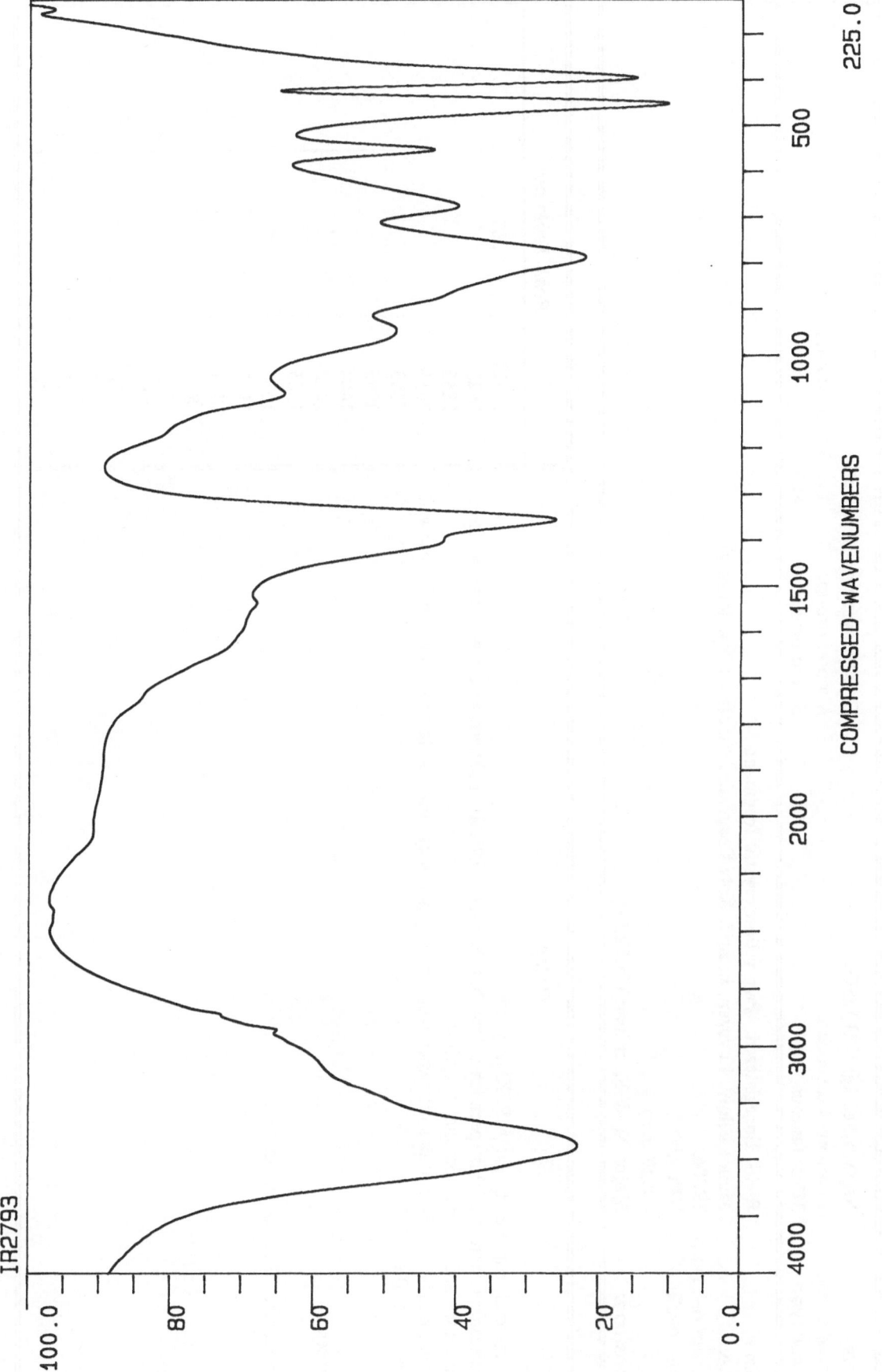

IR2793

% TRANSMITTANCE

COMPRESSED-WAVENUMBERS

MANASSEITE

MANGANOTYCHITE

Formula:	**Na$_6$(Mn,Fe,Mg)$_2$(CO$_3$)$_4$(SO$_4$)**
Chemical class:	**Compound carbonate**
Chemical type:	**Miscellaneous**

Crystal system:	**Cubic**
Mineral group:	**Tychite (susannite)**
Space group:	**Fd3**

Specimen:	**RMS. Unregistered. Pale yellow crystal fragments.**
Source:	**Mount Alluiv, Lovozero massif, Kola Peninsula, Russia.** (Type locality).
Spectrum ref. no.:	**IR3062**
Sample medium:	**KBr disk**
XRD:	**9062F (std)**
Composition:	**Major Na & S, minor Mn,Fe,Mg**

Peak Table cm^{-1}

[3418]	**301**
2923	
2855	
2611	
2503	
1795	
1442	
1420	
1385	
1176	
1111	
874	
829	
799	
710	
631	
515	
467	
351	

Notes

A new mineral, supplied by Dr A.P. Khomyakov.
The manganese content of this specimen is low, but the x-ray diffraction pattern agrees with that of the type specimen and differs from that of tychite.
The spectrum shows slight peak shifts when compared to that of tychite and lacks the three well-resolved peaks below 500 cm^{-1}.

References:

1. Khomyakov A.P. et al., in press.

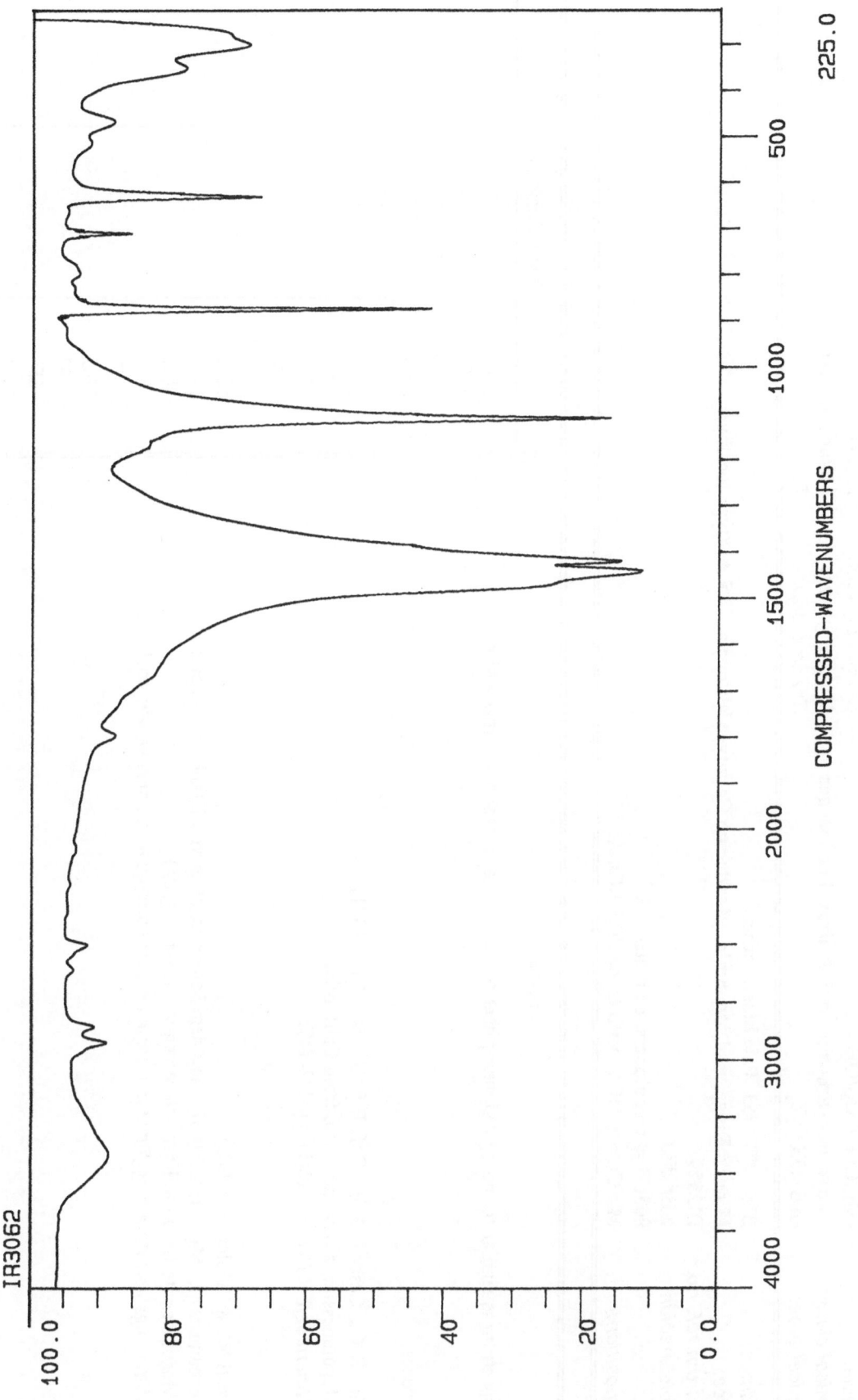

IR3062

% TRANSMITTANCE

COMPRESSED-WAVENUMBERS

MANGANOTYCHITE

MCGUINNESSITE

Formula:	$(Mg,Cu)_2(CO_3)(OH)_2$
Chemical class:	Anhydrous carbonate with hydroxyl or halogen
Chemical type:	$(AB)_5(XO_3)_2Z_q$
Crystal system:	Monoclinic
Mineral group:	Malachite (rosasite)
Space group:	$P2_1/a$

Specimen:	BM 1977, 463 Pale blue coating.
Source:	Miner's Ridge, Red Mountain, Mendocino County, California, U.S.A. (type locality).
Spectrum ref. no.:	IR2829
Sample medium:	KBr disk
XRD:	8080F = mcguinnessite or near
Composition:	Mg:Cu ≈ 1·1:0·9 with trace Si,Fe,Ca,Al

Peak Table cm⁻¹

3546	533
3409	499
3316	**424**
2921	**383**
2570	330
2421	269
2059	
1792	
1548	
1432	
1392	
1101	
1050	
855	
834	
742	
707	
656	
561	

Notes

The spectrum is similar to those of glaukosphaerite, kolwezite, rosasite and malachite

References:

1. Erd R.C., Cesbron F.P., Goff F.E. & Clark J.R. (1981)
    Mcguinnessite, a new carbonate from California.
    Mineralogical Record, **12**(3), pp.143-147.

2. Postl W. & Golob P. (1981)
    Mcguinnessit, $(Mg,Cu)_2CO_3(OH)_2$ aus dem Serpentingebiet von Kraubath, Steiermark.
    (Mcguinnessite from the Kraubath serpentine massif, Styria).
    Mitteilungsblatt Abteilung für Mineralogie am Landesmuseum Joanneum, **49**, pp.15-21.

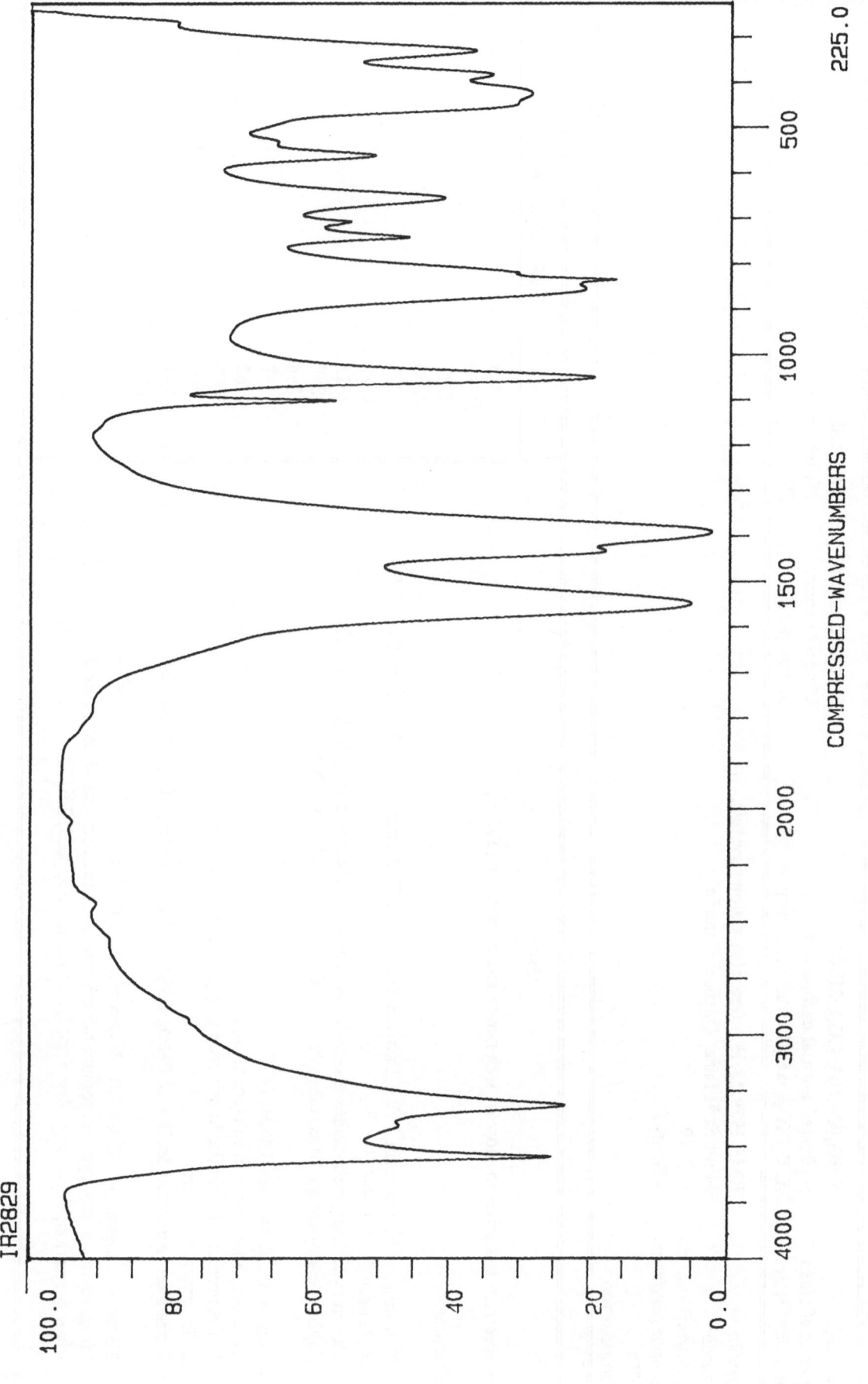

IR2829

% TRANSMITTANCE

100.0   80   60   40   20   0.0

4000   3000   2000   1500   1000   500   225.0

COMPRESSED-WAVENUMBERS

MCGUINNESSITE

MCKELVEYITE-(Y)

Formula:	$NaBa_3(Ca,U)Y(CO_3)_6 \cdot 3H_2O$
Chemical class:	Hydrated normal carbonates
Chemical type:	$A_mB_n(XO_3)_p \cdot xH_2O$ where (m+n):p = 1:1

Crystal system:	Trigonal
Mineral group:	Mckelveyite
Space group:	P$\bar{3}$

Specimen:	RMS 1979.25.15. Greenish yellow crystals.
Source:	Mont St Hilaire, Quebec, Canada.
Spectrum ref. no.:	IR2949
Sample medium:	KBr disk
XRD:	
Composition:	

Peak Table cm^{-1}

3413
3270
2927
2850
1797
1676
1515
1390
1364
1139
1062
1017
856
723
694
648
603
425

Notes

The spectrum is similar to those of **weloganite** and **donnayite-(Y)**.

References:

1. Voloshin A.V., Subbotin V.V., Yavoventchuk V.N., Pakhomovsky Y.A., Menshikov Y.P. & Zaitsev A.N. (1990)
   Mckelveyite from carbonatites and hydrothermalites of alkaline rocks, Kola Peninsula.
   Zapiski Vsesoyuznogo Mineralogicheskogo Obshchestva, **119**, pp.76-86.

2. Donnay G. & Donnay J.D.H. (1971)
   Ewaldite, a new barium calcium carbonate.
   1. Occurrence of ewaldite in syntactic intergrowth with mackelveyite.
   2. Its crystal structure.
   Tschermaks Mineralogische und Petrographische Mitteilungen, **15**, pp.185-212.

2. Milton C., Ingram B., Clark J.R. & Dwornik E.J. (1965)
   Mckelveyite, a new hydrous sodium barium rare earth uranium carbonate mineral from the Green River formation, Wyoming. *American Mineralogist,* **50**, pp.593-612.

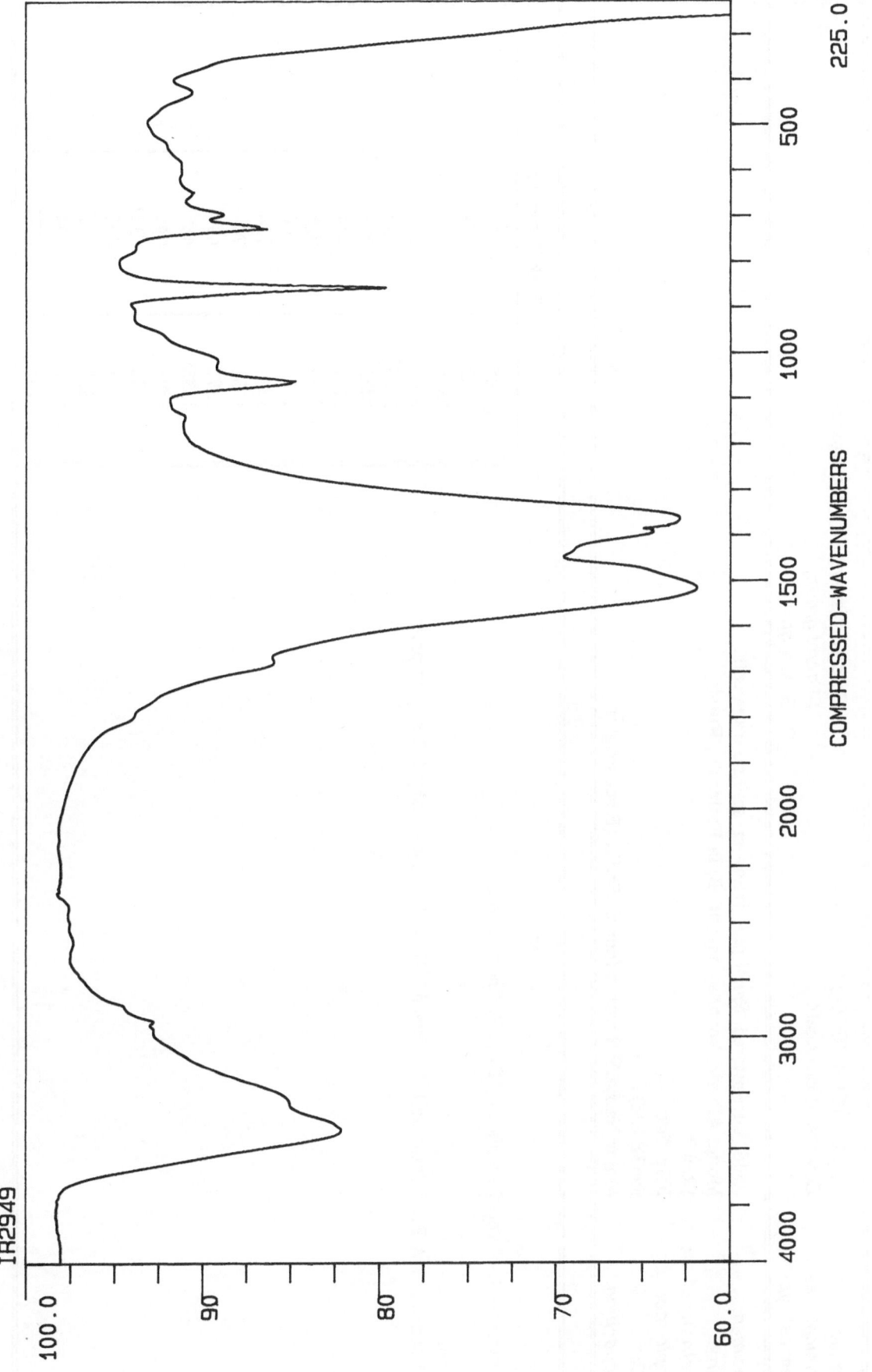

IR2949

% TRANSMITTANCE

COMPRESSED-WAVENUMBERS

MCKELVEYITE- (Y)

4000   3000   2000   1500   1000   500   225.0

100.0   90   80   70   60.0

MINEEVITE-(Y)

Formula:	$Na_{25}BaY_2(CO_3)_{11}(HCO_3)_4(SO_4)_2F_2Cl$	Crystal system:	Hexagonal.
Chemical class:	Compound carbonate	Mineral group:	
Chemical type:		Space group:	P6₃/m

Specimen:	RMS, unregistered. Small pale yellow crystalline fragments.
Source:	Mount Alluaiv, Lovozero massif, Kola Peninsula, Russia.
Spectrum ref. no.:	IR3068
Sample medium:	KBr disk
XRD:	9063F (std)
Composition:	Major Na,Ba,Y,S with minor Cl,Sr,Si, (F not sought).

Peak Table cm⁻¹

[3408]	1122
2923	1071
2855	1055
2604	914
2555	892
2433	877
2147	865
2114	815
1781	766
1744	724
1631	704
1592	693
1535	688
1506	645
1433	632?
1392?	465
1375	369
1361	304
1147	257?

Notes

A new mineral. Material supplied by Dr A.P. Khomyakov.

References:

1. Khomyakov A.P., Polezhaeva L.I., Yamnova N.A. & Pusharovsky D.Yu., in press.

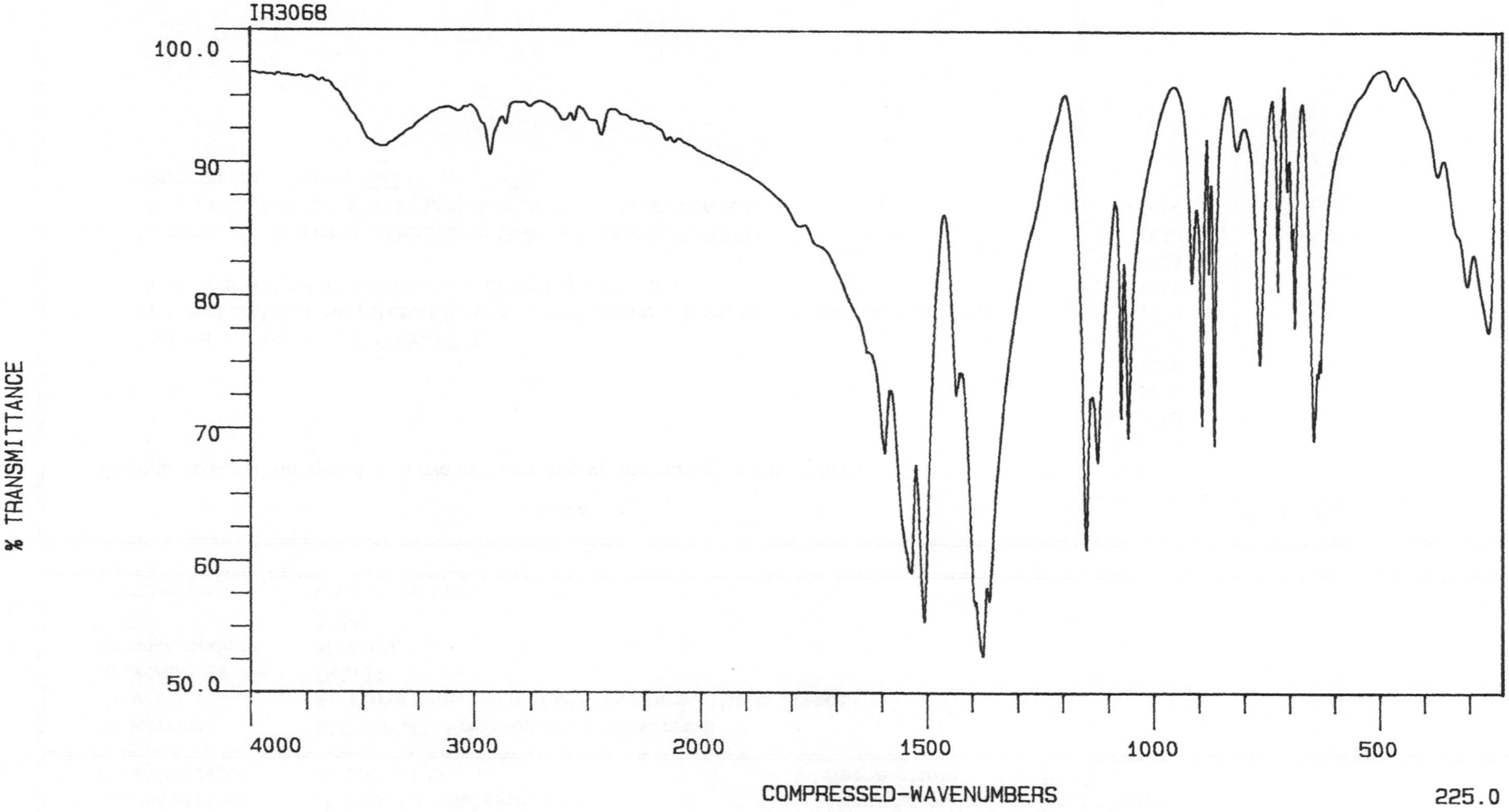

IR3068

% TRANSMITTANCE

100.0   90   80   70   60   50.0

4000   3000   2000   1500   1000   500   225.0

COMPRESSED-WAVENUMBERS

MINEEVITE

MONOHYDROCALCITE

Formula:	$CaCO_3 \cdot H_2O$
Chemical class:	Hydrated normal carbonate
Chemical type:	$A(XO_3) \cdot xH_2O$

Crystal system:	**Trigonal**
Mineral group:	**Nesquehonite**
Space group:	**P3₁12**

Specimen:	BM 1979,47 Bright blue crystalline crust.
Source	St. Pierre mine, Saint-Marie-aux-Mines, Alsace, France.
Spectrum ref. no.:	IR2753
Sample medium:	KBr disk
XRD:	20385
Composition:	Ca only (see notes).

Notes

The blue colour of the specimen is due to a very thin surface coating of cuproadamite.

References:

1. Řidkosil T., Sejkora J. & Ondrus P. (1991)
    Monohydrocalcite from polymetallic vein of the Vrančice deposit, near Příbram, Czechoslovakia.
    Neues Jahrbuch für Mineralogie, Monatshefte, pp.289–95.

2. Catherine H., Skinner W., Osbaldiston G.W. & Wilner A.N. (1977)
    Monohydrocalcite in a guinea pig bladder stone, a novel occurrence.
    American Mineralogist, **62**(3,4), pp.273–77.

Peak Table cm⁻¹

3319	234?
3232	
2553	
2472	
2270	
2135	
1789	
1767	
1703	
1484	
1409	
1069	
873	
765	
726	
700	
674	
590	
284	

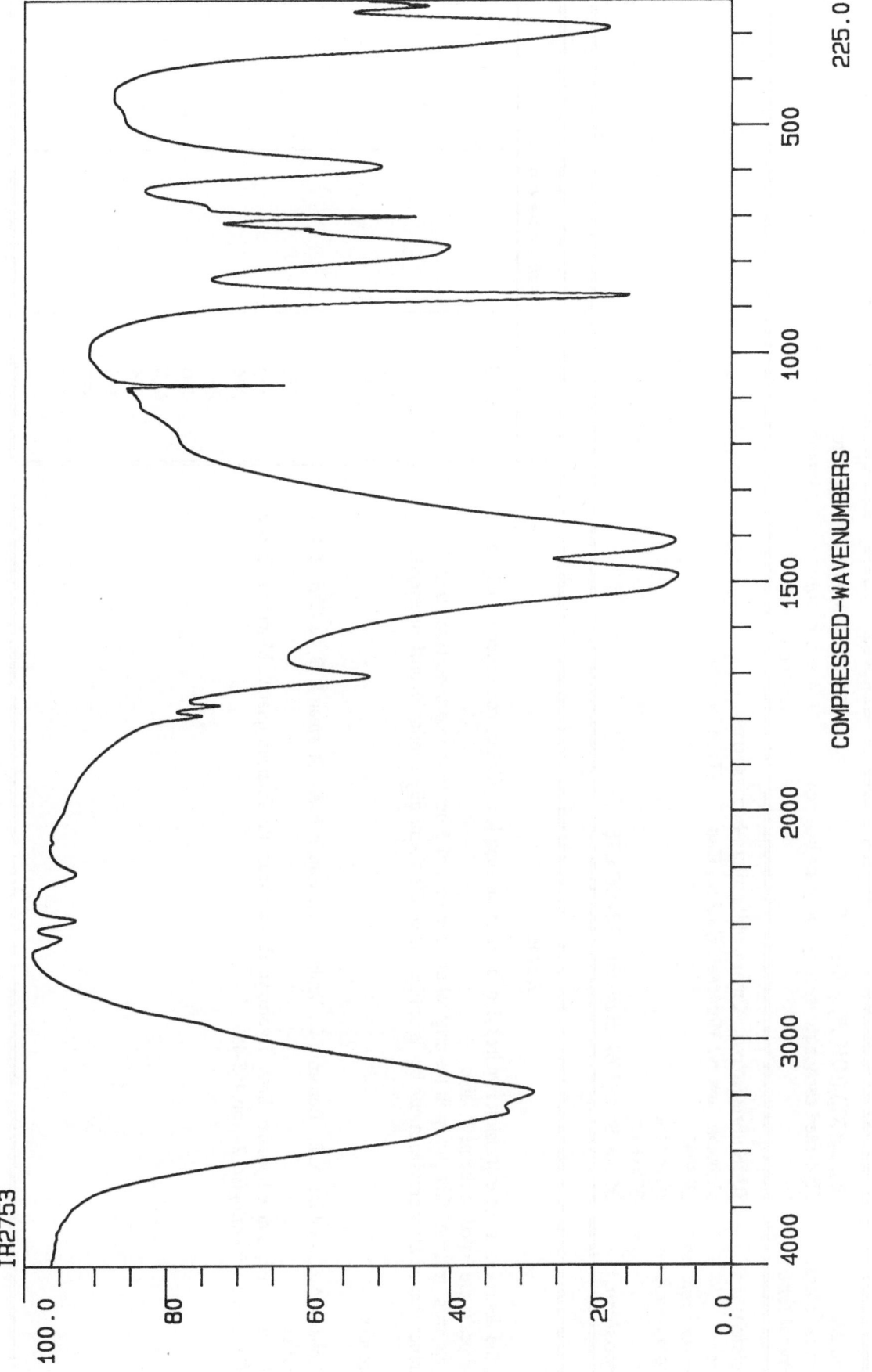

IR2753

% TRANSMITTANCE

COMPRESSED-WAVENUMBERS

MONOHYDROCALCITE

MONTROYALITE

Formula:	$Sr_4Al_8(CO_3)_3[(OH),Fl_{26}]\cdot10\text{-}11H_2O$
Chemical class:	Hydrated carbonate with hydroxyl or halogen
Chemical type:	
Crystal system:	Triclinic ?
Mineral group:	Alumohydrocalcite ?
Space group:	?

Specimen:	RMS unregistered. Cream/white micro-hemispheres.
Source:	Francon quarry, Montreal, Quebec, Canada. (Type locality).
Spectrum ref. no.:	IR3003
Sample medium:	KBr disk
XRD:	8791F
Composition:	Major Sr and Al plus trace S,Na,Ca,Fe

Notes

Note the degree of scale expansion required due to the small sample available, consequently only the major peaks are listed in the peak table.

The spectrum matches that shown in the original description, ref.1 but with improved resolution. It resembles that of **alumohydrocalcite** but is unlike that of the chemically similar **strontiodresserite**.

References:

1. Roberts A.C., Sabina A.P., Bonardi M., Jambor J.L., Ramik R.A., Sturman B.D. & Carr M.J. (1986) Montroyalite, a new hydrated Sr-Al Hydroxycarbonate from the Francon quarry, Montreal, Quebec. *Canadian Mineralogist*, **24**, pp.455-459.

Peak Table cm^{-1}

3518		
3480		
3360		
3214		
1662		
1547		
1456		
1390		
1051		
990		
896		
833		
750		
610		
556		
486		
372		
304		

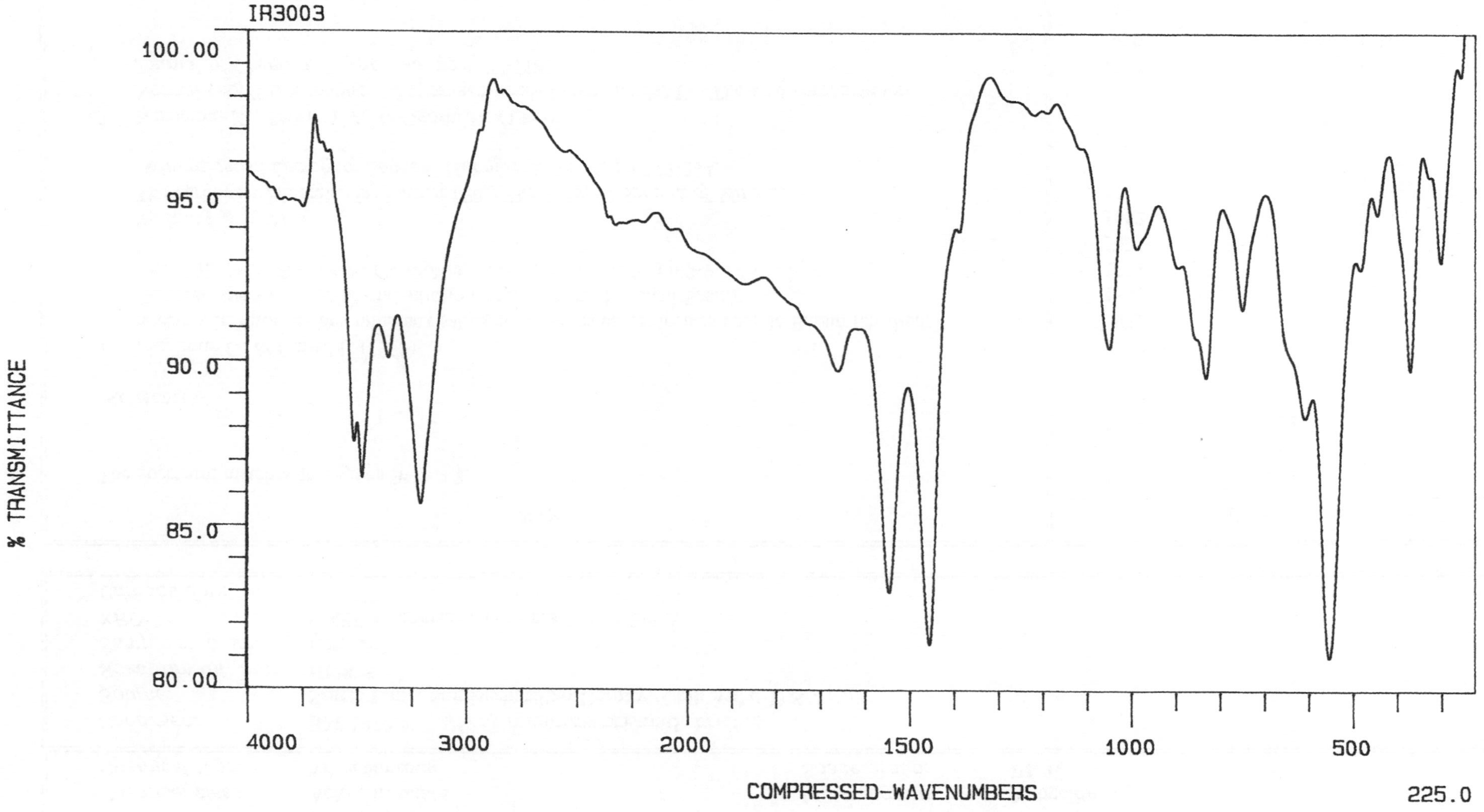

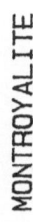

IR3003

% TRANSMITTANCE

100.00   95.0   90.0   85.0   80.00

4000   3000   2000   1500   1000   500   225.0

COMPRESSED-WAVENUMBERS

MONTROYALITE

NAHCOLITE

Formula:	NaHCO$_3$	Crystal system:	Monoclinic
Chemical class:	Acid carbonate	Mineral group:	Nahcolite
Chemical type:	Miscellaneous	Space group:	P2$_1$/b

Specimen:	BM 1934,47 Glassy colourless prismatic crystals.
Source:	Searls Lake, San Bernardino County, California, U.S.A.
Spectrum ref. no.:	IR2875
Sample medium:	KBr disk
XRD:	8148F = nahcolite (+ weak line at 3·8 Å)
Composition:	

Peak Table cm^{-1}

3438	
3075	
2919	
2544	814
2268	697
2043	658
1922	
1842	
1731	
1696	
1662	
1618	
1453	
1499	
1308	
1047	
1033	
999	
837	

Notes

The spectrum matches that given in ref. 2.

References:

1. Maglione G. & Carn M. (1975)
    Spectres infrarouges des mineraux salins et des silicates neoformes dans le Bassin tchadien.
    (Infrared spectra of saline and silicate minerals from the Chad Basin).
    Fr., Off. Rech. Sci. Tech. Outre Mer, Cah., Ser. Geol. 7, pp.3-9

2. White W.B. (1974)
    The carbonate minerals. *In:* Farmer (Ed.) *The Infrared Spectra of Minerals.*
    Mineralogical Society of London, Monograph No. 4, pp.227-284.

3. Nakamoto K., Sarma Y.A. & Ogoshi H. (1965)
    Normal coordinate analyses of hydrogen-bonded compounds. IV. The acid carbonate ion.
    Journal of Chemical Physics, **43**, pp.1177-1181.

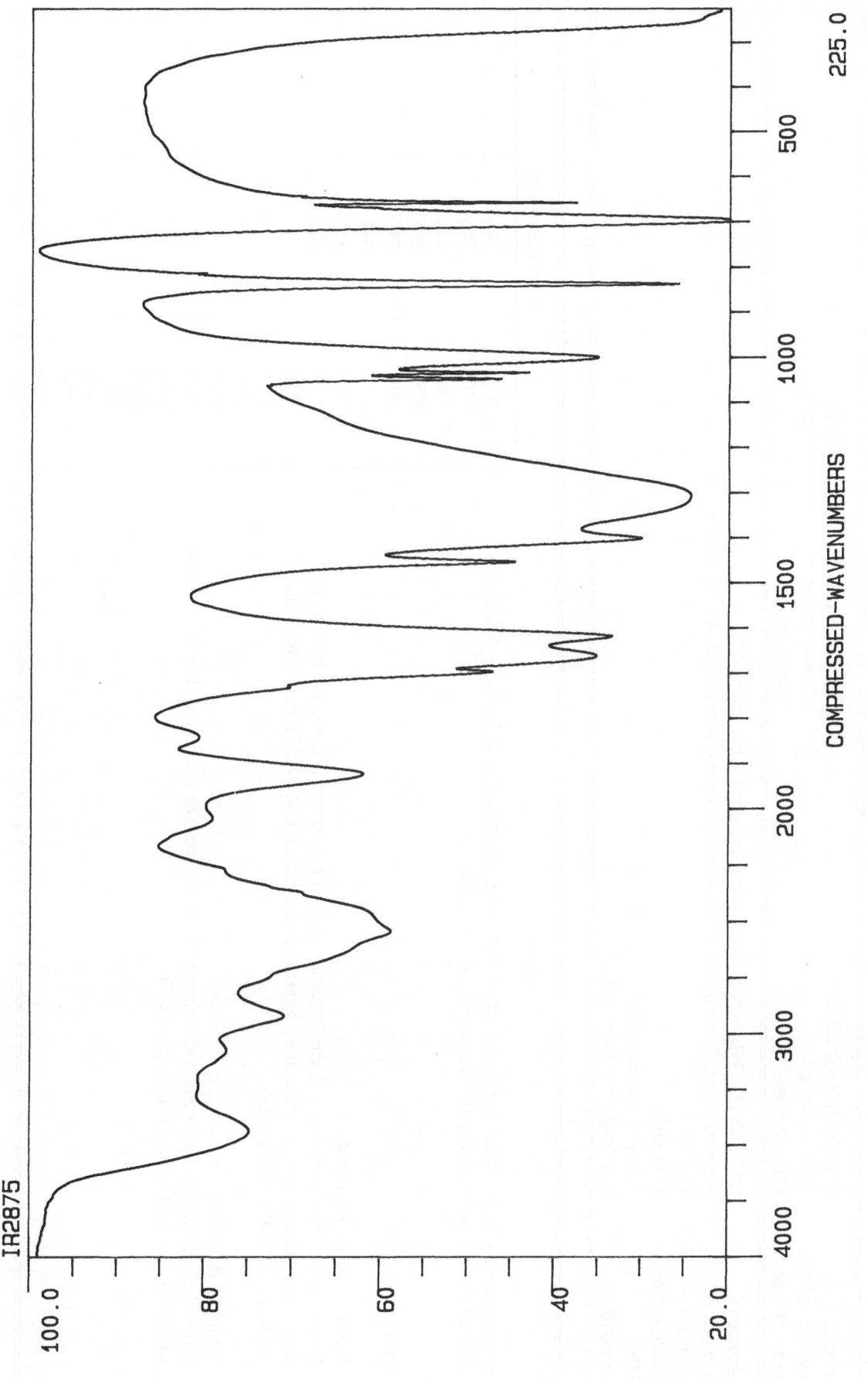

% TRANSMITTANCE

COMPRESSED-WAVENUMBERS

NAHCOLITE

IR2875

NESQUEHONITE

Formula:	**Mg(HCO$_3$)(OH)·2H$_2$O**
Chemical class:	**Acid carbonate**
Chemical type:	**Miscellaneous**

Crystal system:	**Monoclinic**
Mineral group:	**Nesquehonite**
Space group:	**P2$_1$/n**

Specimen:	**BM 1921,53 White radiating crystalline crust with anthracite.**
Source:	Nesquehoning, Lansford, Carbon County, Pennsylvania, U.S.A. (Type locality)
Spectrum ref. no.:	IR2767
Sample medium:	KBr disk
XRD:	7829F = nesquehonite (+ giorgiosite).
Composition:	Mg with trace Si,Na,Ca,Cl

Peak Table cm^{-1}

3606	**748**
3563	**706**
3443	**662**
3358	627
3297	**500**
2620	**443**
2511	**390**
2194	**272**
1678	
1636	
1591	
1531	
1472	
1442	
1100	
1029	
932	
854	
794	

Notes

The specimen may be a mixture due to conversion to/from giorgiosite Mg$_5$(CO$_3$)$_4$(OH)$_2$·5H$_2$O but the spectrum is close to that of nesquehonite shown in Farmer.

References:

1. Suzuki J. & Ito M. (1974)
Nesquehonite from Yoshikawa, Aichi Prefecure, Japan: occurrence and thermal behaviour.
Journal of the Japanese Association of Mineralogists, Petrologists and Econonomic Geologists, **69**(8), pp.275-284. (In English)

2. White W.B. (1971)
Infrared characterization of water and hydroxyl ion in the basic magnesium carbonate minerals.
American Mineralalogist, **56**(1,2), pp. 46-53.

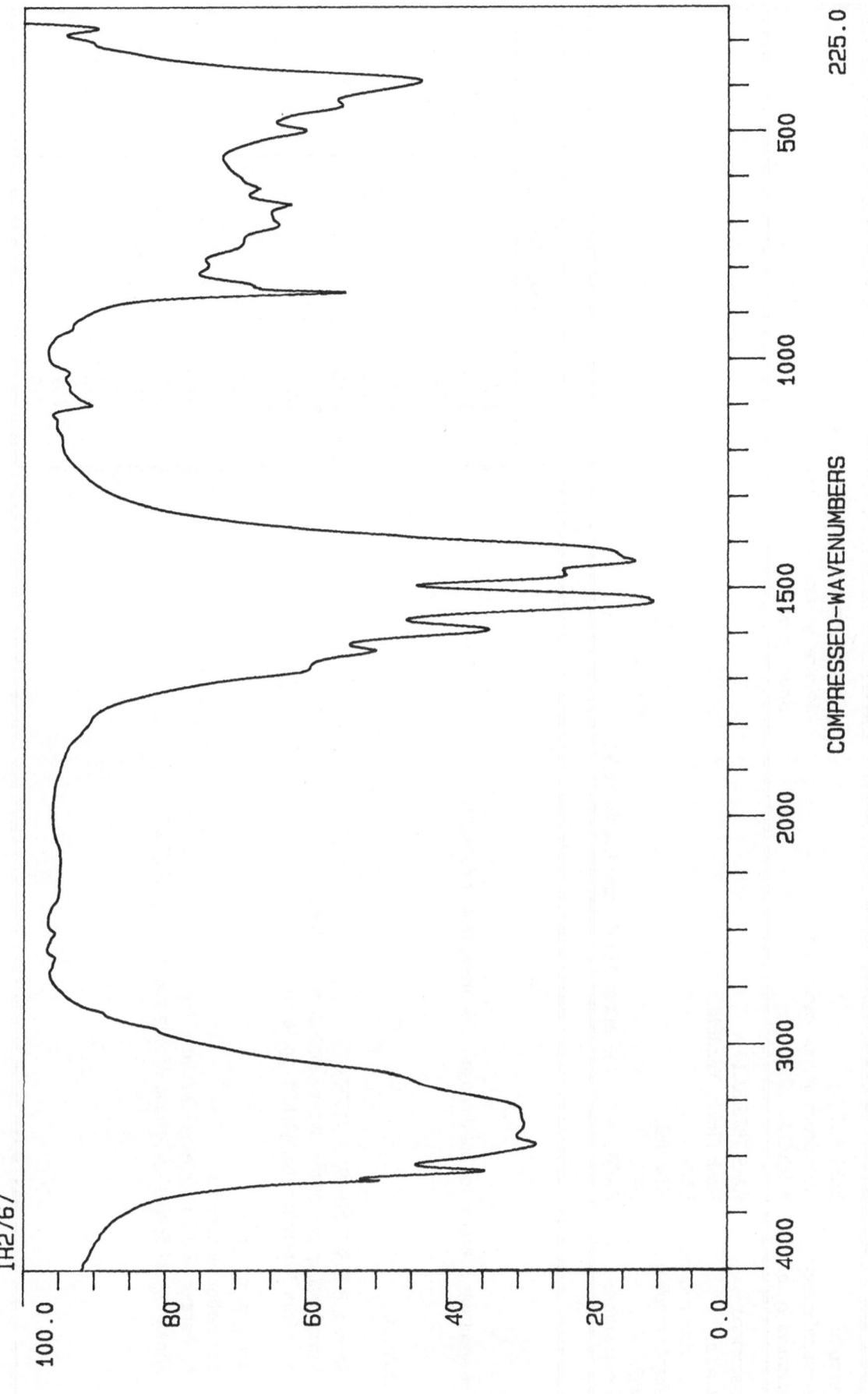

IR2767

% TRANSMITTANCE

100.0   80   60   40   20   0.0

4000   3000   2000   1500   1000   500   225.0

COMPRESSED-WAVENUMBERS

NESQUEHONITE

NORSETHITE

Crystal system:	Trigonal	
Mineral group:	Dolomite	
Space group:	R32	

Formula:	BaMg(CO$_3$)$_2$
Chemical class:	Anhydrous normal carbonate
Chemical type:	AB(XO$_3$)$_3$
Specimen:	RMS 1976.33.14
Source:	Rosh Pinah, Namibia.
Spectrum ref. no.:	IR2911
Sample medium:	KBr disk
XRD:	3902
Composition:	Ba:Mg = 0·7:1 + minor Mn & trace Ca,Na,Fe,Sr

Peak Table cm^{-1}

[3431]
2963
2924
2873
2649
2524
2362
2335
1802
1448
1126
1116
880
853?
713
702
636
609
349

Notes

The spectrum is similar to, but distinguishable from, that of **dolomite**

References:

1. Scheetz B.E & White W.B. (1977)
    Vibrational spectra of the alkaline earth double carbonates.
    American Mineralalogist. 62(1,2), pp.36-50.

2. White W.B. (1974)
    The carbonate minerals.
    In: Farmer (Ed) *The Infrared Spectra of Minerals.*
    Mineralogical Society of London, Monograph No. 4, pp.227-284.

NORTHUPITE

Formula:	Na$_3$Mg(CO$_3$)$_2$Cl
Chemical class:	Anhydrous carbonate with hydroxyl or halogen
Chemical type:	(AB)$_2$(XO$_3$)Z$_q$

Crystal system:	Cubic
Mineral group:	Bastnäsite
Space group:	Fd$\bar{3}$

Specimen:	BM 1905,246 Isolated grey/white octahedra.
Source:	Borax Lake, San Bernardino County, California, U.S.A.
Spectrum ref. no.:	IR2789
Sample medium:	KBr disk
XRD:	7884F = northupite
Composition:	Na:Mg:Cl $\approx$ 3:1·2:1 with trace Si.

Notes

See Farmer for discussion of the spectrum.
The three strong peaks below 500 cm^{-1} are beyond the range shown by Adler and Kerr (1963b).

References:

1. Maglione G. & Carn M. (1975)
Spectres infrarouges des mineraux salins et des silicates neoformes dans le Bassin tchadien
(Infrared spectra of saline and silicate minerals from the Chad Basin).
Fr., Off. Rech. Sci. Tech. Outre Mer, Cah., Ser. Geol. 7, pp.3-9.

2. Pratt J. H. (1896)
On northupite; pirssonite, a new mineral; gaylussite and hanksite from Borax Lake, San Bernardino Co., California.
American Journal of Science 4(2) pp.123-35.
Zeitschrift für Kristallographie,27, pp.416-29 and (1901) *Yale Bicen. Pub.Contr. Miner.*, pp.261-74.

Peak Table cm^{-1}

[3433]	
2923	336
2857	273
2659	
2558	
2360	
1819	
1629	
1463	
1449	
1168	
1090	
1019	
880	
856	
798	
713	
672	
397	

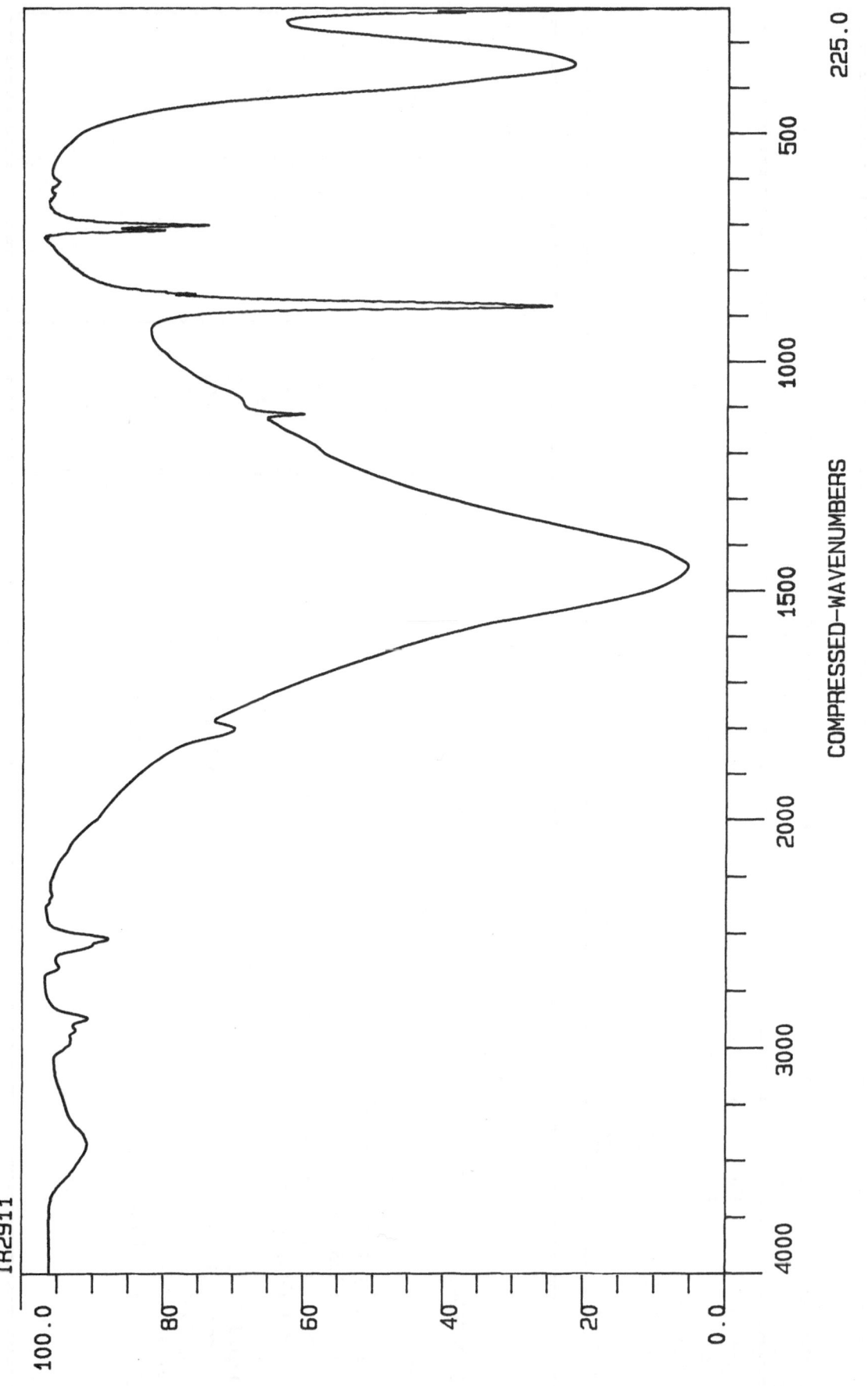

IR2911

% TRANSMITTANCE

100.0   80   60   40   20   0.0

4000   3000   2000   1500   1000   500   225.0

COMPRESSED-WAVENUMBERS

NORSETHITE

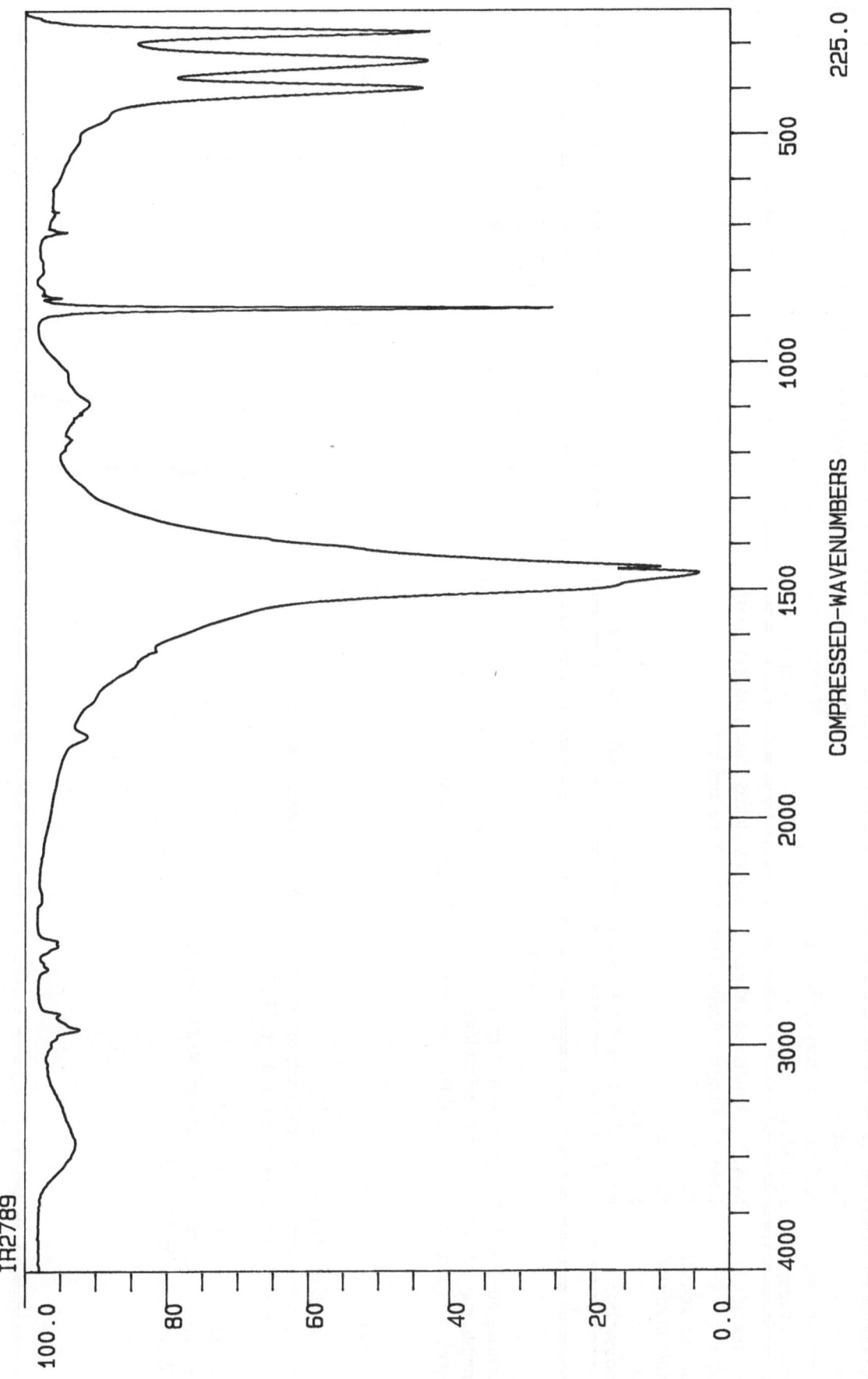

IR2789

% TRANSMITTANCE

COMPRESSED-WAVENUMBERS

NORTHUPITE

NYEREREITE

Formula:	Na₂Ca(CO₃)₂
Chemical class:	Anhydrous normal carbonate
Chemical type:	A₂B(XO₃)₂

Crystal system:	Orthorhombic, pseudo-hexagonal
Mineral group:	Eitelite
Space group:	Cmc2₁

Specimen:	M 42171 Colourless tabular fragments picked from powdery debris
Source:	Oldoinyo Lengai volcano, Tanzania. (Type locality)
Spectrum ref. no.:	IR2885
Sample medium:	KBr disk
XRD:	8200F (std)
Composition:	Na:Ca:K = 1·6:1·0:0·4 with minor Sr, Mg, S and trace P, Ba

Peak Table cm⁻¹

[3441]	258
2983	
2899	
2599	
2528	
2337	
1795	
1468	
1186	
1144	
1108	
1079	
1010	
873	
710	
689	
648	
622	
573	

Notes

Specimen from Royal Ontario Museum, Canada

Nyerereite may be identical to natrofairchildite

Compare the spectrum with those of the calcite and dolomite group minerals.

References:

1. McKie D. & Frankis E.J. (1977)
    Nyerereite; a new volcanic carbonate mineral from Oldoinyo Lengai, Tanzania.
    Zeitschrift für Kristallographie, **145**, pp.73-95.

2. Frankis E.J & McKie D. (1973)
    Subsolidus Relations in the System Na₂CO₃CaCO₃·H₂O
    Nature, **246**, pp.124-126.

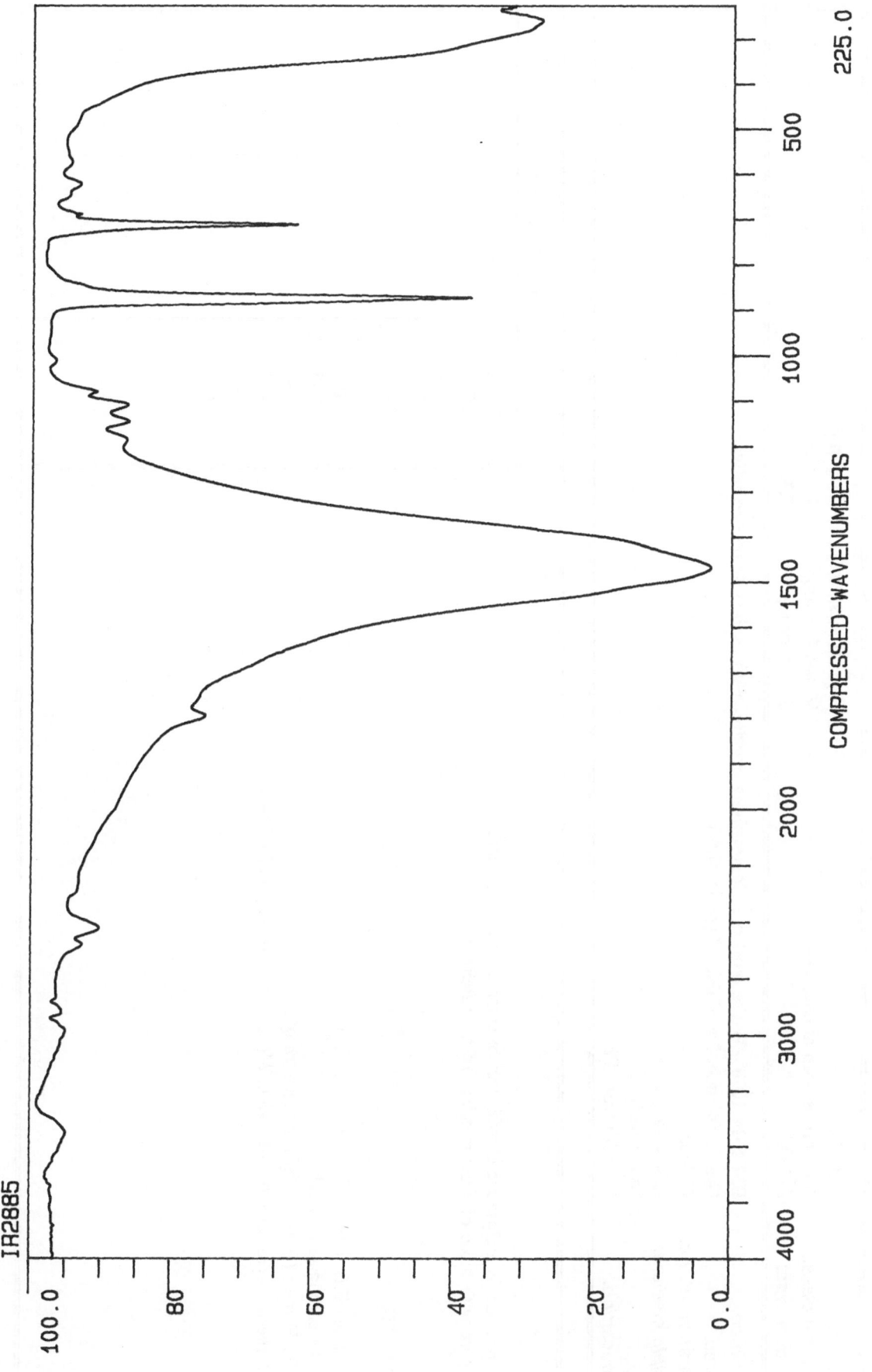

IR2885

% TRANSMITTANCE

100.0

80

60

40

20

0.0

4000

3000

2000

1500

1000

500

225.0

COMPRESSED-WAVENUMBERS

NYEREREITE

OTAVITE

Formula:	CdCO$_3$
Chemical class:	Anhydrous normal carbonate
Chemical type:	A(XO$_3$)

Crystal system:	Trigonal
Mineral group:	Calcite
Space group:	R3̄c

Specimen:	BM 1914,1070 White crust pseudomorphing cuprite, with malachite on cerussite.
Source:	Otavi, Tsumeb, Namibia. (Type locality).
Spectrum ref. no.:	IR2653
Sample medium:	KBr disk
XRD:	7401F (std)
Composition:	Cd with trace Pb

Peak Table cm^{-1}

[3406]
2923
2854
2797
2468
1800
1407
860
834
723
465
293

Notes

The spectrum is a close match with that of synthetic cadmium carbonate.
Compare with those of other members of the **calcite** group.

References:

1. White W.B. (1974)
    The carbonate minerals.
    In: Farmer (Ed) *The Infrared Spectra of Minerals.*
    Mineralogical Society of London, Monograph No. 4, pp.227-284.

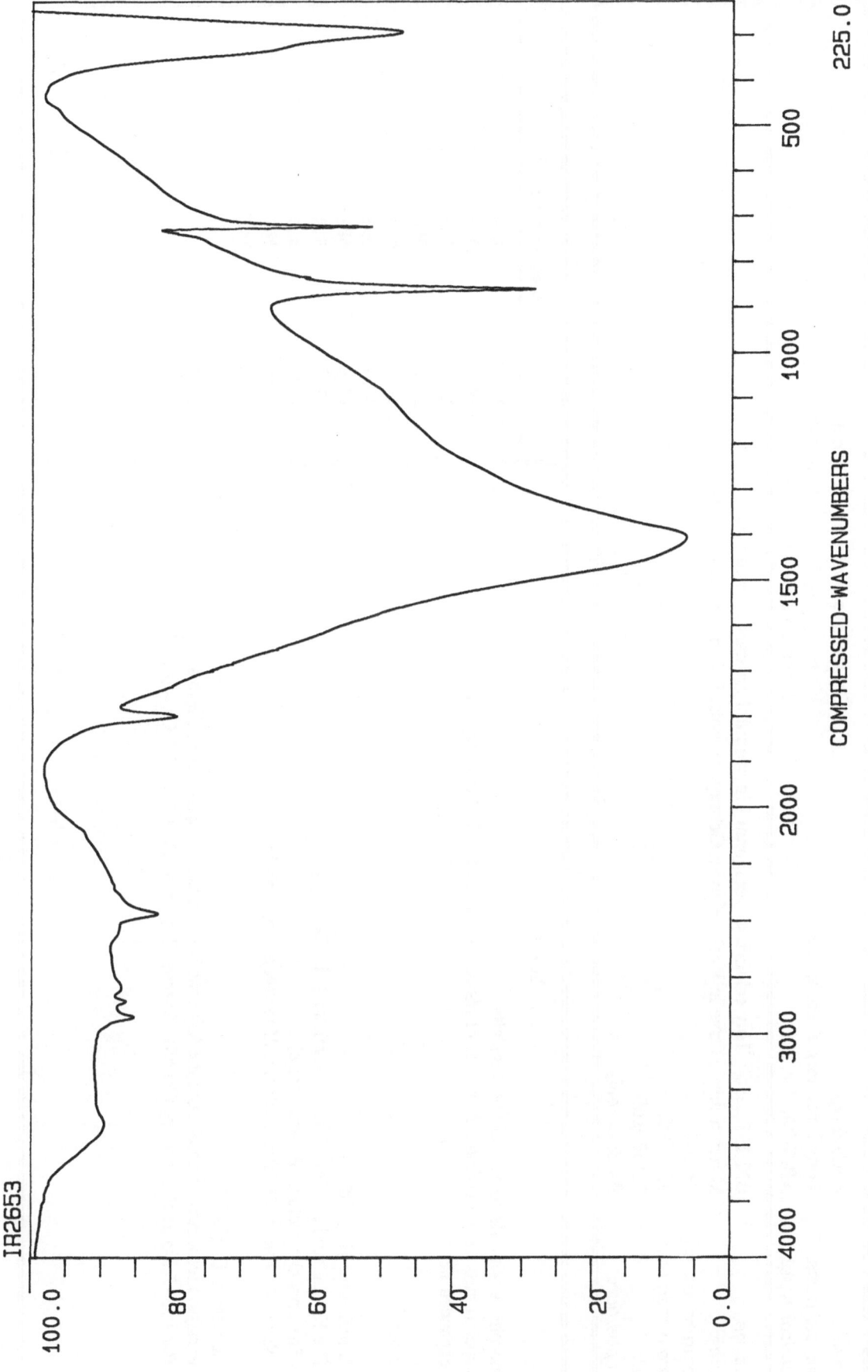

IR2653

% TRANSMITTANCE

100.0
80
60
40
20
0.0

4000   3000   2000   1500   1000   500   225.0

COMPRESSED-WAVENUMBERS

OTAVITE

PARALSTONITE

Formula:	BaCa(CO$_3$)$_2$
Chemical class:	Anhydrous normal carbonate
Chemical type:	AB(XO$_3$)$_3$
Crystal system:	Trigonal
Mineral group:	Aragonite
Space group:	P321

Specimen:	BM 1972,254 Tiny colourless hexagonal bipyramidal crystals on witherite.
Source:	Minerva mine, Cave-in-Rock, Hardin County, Illinois, U.S.A. (Type locality).
Spectrum ref. no.:	IR2872
Sample medium:	KBr disk
XRD:	8146F (std)
Composition:	Ba & Ca only

Peak Table cm^{-1}

[3341]	855
2923	838
2855	708
2570	**700**
2492	693
2464	**691**
1765	682
1755	518
1507	**466**
1486	**297**
1448	242?
1409	
1183	
1085	
1064	
903	
898	
893	
862	

Notes

Trimorphous with alstonite and barytocalcite.
The spectrum is distinguishable from that of **alstonite** only in the 700 cm^{-1} region.
See expanded detail.

References:

1. Effenberger H. (1980)
    Die Kristallstruktur des Minerals Paralstonit, BaCa(CO$_3$)$_2$
    (The crystal structure of paralstonite).
    Neues Jahrbuch für Mineralogie, Monatshefte, pp.353-63.

2. Roberts A.C. (1979)
    Paralstonite; a new mineral from the Minerva No.1 mine, Cave in Rock, Illinois.
    Papers, Geological Survey of Canada, Current Research, Part C, 79(1,C), pp.99-100.

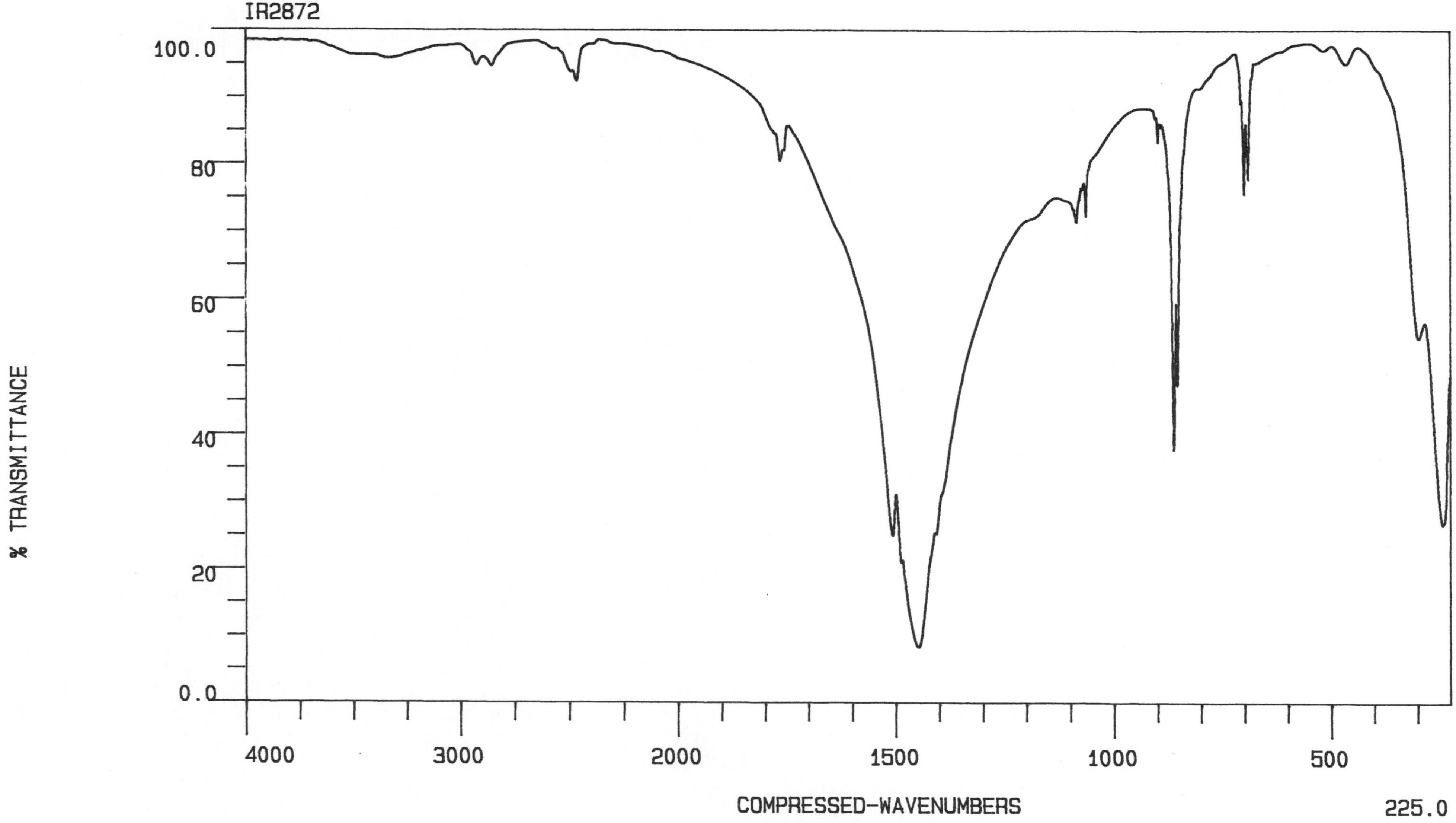

IR2872

COMPRESSED-WAVENUMBERS

% TRANSMITTANCE

PARALSTONITE

100.0 80 60 40 20 0.0

4000 3000 2000 1500 1000 500 225.0

% TRANSMITTANCE

WAVENUMBERS

IR2872

PARALSTONITE   [expanded detail]

PARISITE-(Ce)

Formula:	$Ca(Ce,La)_2(CO_3)_3F_2$	*Crystal system:* Trigonal
Chemical class:	Anhydrous carbonate with hydroxyl or halogen	*Mineral group:* Bastnäsite
Chemical type:	$(AB)(XO_3)Z_q$	*Space group:* R3

Specimen: BM 1924,854 Brown striated barrel-shaped hexagonal crystals.
Source: Narsarssuak, Julianehaab, Greenland.
Spectrum ref. no.: IR2892
Sample medium: KBr disk
XRD: 11596 = parisite
Composition: Ca:Ce:La:Nd = 1·0:0·8:0·4:0·4 with trace Fe,Pr,Y,Th

Peak Table cm^{-1}

[3483]		
3157		
2834		
2493		
2348		
1818		
1749		
1457		
[1412]		
1076		
1009		
977		
871		
739		
684		
361		
318		
267		
250		

Notes

The spectrum matches that in Suhner (5-37A) with the exception of those peaks due to absorbed water.

References:

1. Akhmanova M.N. & Orlova L.P. (1966)
    Investigation of rare-earth carbonates by infra-red spectroscopy.
    Geokhimiya, No.5, pp.571-578.
    Translated in: *Geochemistry International.*, **3**(3), pp. 444-451.

2. Adler H.H. & Kerr P.F. (1963)
    Infrared spectra, symmetry and structure relations of some carbonate minerals.
    American Mineralogist, **48**, pp.839-853.

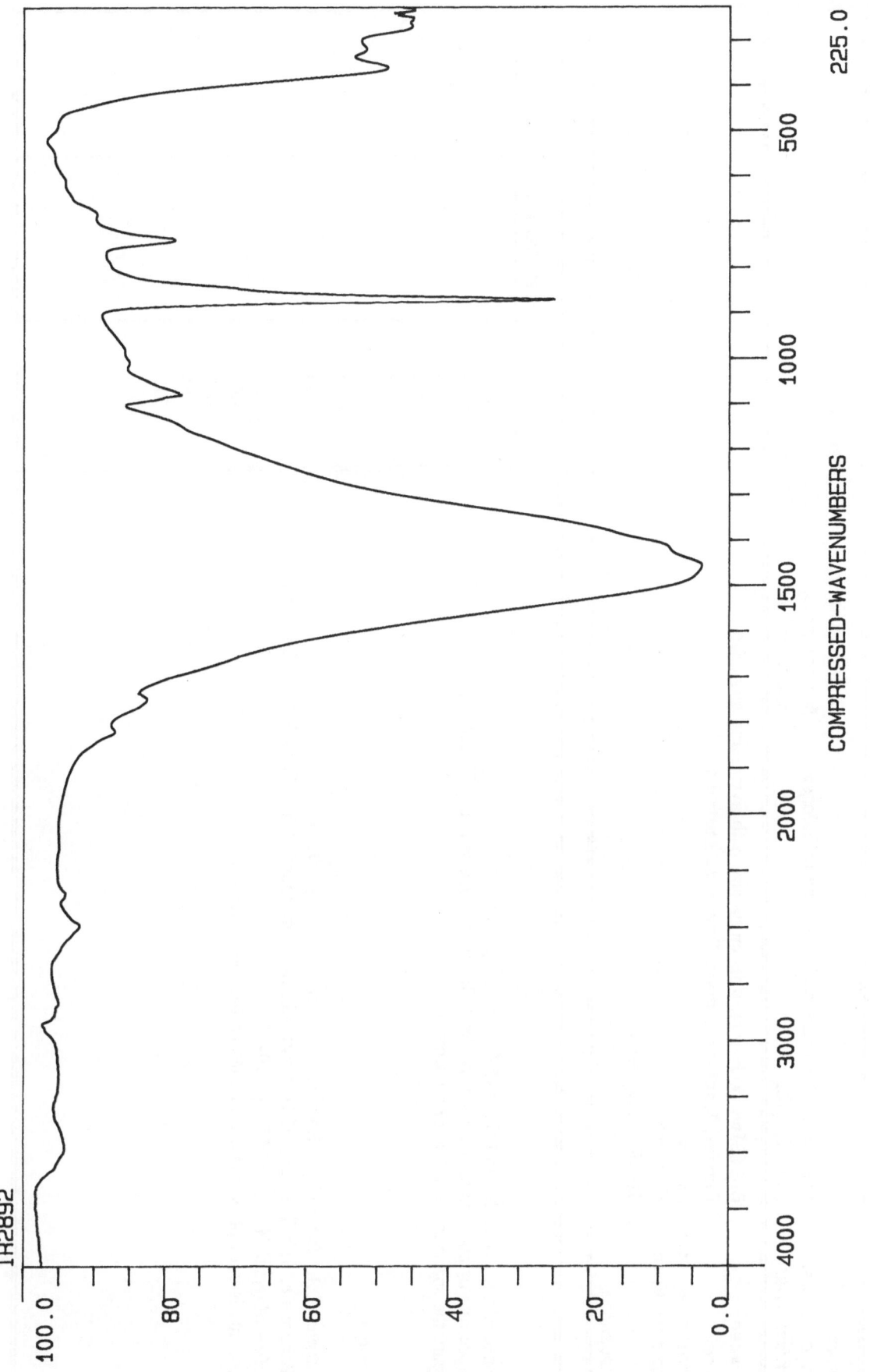

IR2892

% TRANSMITTANCE

COMPRESSED-WAVENUMBERS

PARISITE- (Ce)

PHOSGENITE

Formula:	**Pb$_2$(CO$_3$)Cl$_2$**
Chemical class:	**Anhydrous carbonate with hydroxyl or halogen**
Chemical type:	**(AB)$_2$(XO$_3$)Z$_q$**

Crystal system:	**Tetragonal**
Mineral group:	**Phosgenite**
Space group:	**P4/mbm**

Specimen:	**BM 85166 Yellow crystals on galena with pyrite and anglesite.**
Source:	**Dundas, Montagu County, Tasmania, Australia.**
Spectrum ref. no.:	**IR2710**
Sample medium:	**KBr disk**
XRD:	**8202F = phosgenite**
Composition:	

Peak Table cm^{-1}

[3422]		
1817		
1710		
1509		
1343		
1128		
1062		
836		
811		
758		
648		
639		
464		
311		

Notes

The spectrum matches that in Suhner (5-25 A), phosgenite.
The spectrum of phosgenite shown in Sadtler (90) has an extra peak at 670 cm^{-1}.
Compare the spectrum with that of **barstowite**.

References:

1. Stanley C.J., Jones G.C., Hart A.D., Keller P. & Lloyd D. (1991)
    Barstowite, 3PbCl$_2$·PbCO$_3$· H$_2$O, a new mineral from Bounds Cliff, St.Endellion, Cornwall.
    Mineralogical Magazine, **55**, pp.119-123.
    (comparison of phosgenite, cerussite and barstowite spectra)

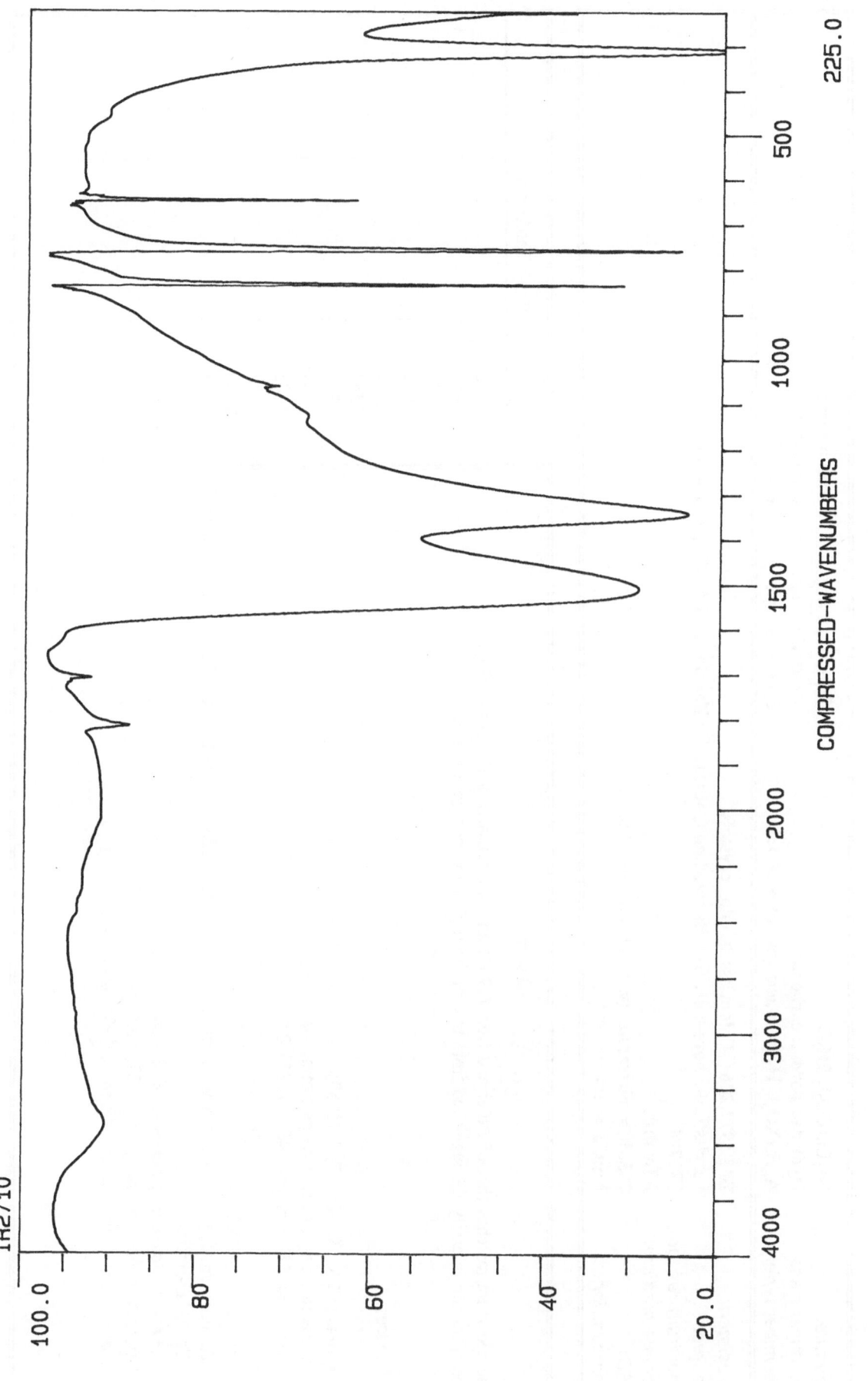

IR2710

% TRANSMITTANCE

100.0

80

60

40

20.0

4000

3000

2000

1500

1000

500

225.0

COMPRESSED-WAVENUMBERS

PHOSGENITE

PIRSSONITE

Crystal system:	**Orthorhombic**
Mineral group:	**Gaylussite**
Space group:	**Fdd2**

Formula:	$Na_2Ca(CO_3)_2 \cdot 2H_2O$
Chemical class:	**Hydrated normal carbonate**
Chemical type:	$A_mB_n(XO_3)_p \cdot H_2O$ where $(m+n){:}p > 1{:}1$

Specimen:	BM 1972,206 Colourless tabular crystals.
Source:	Searles Lake bore-hole, San Bernardino County, California, U.S.A. (Type locality).
Spectrum ref. no.:	IR2784
Sample medium:	KBr disk
XRD:	7856F = pirssonite (with an additional line at 3.3 Å)
Composition:	Na:Ca = 2:1

Peak Table cm^{-1}

3326
3219
3073
2524
2461
2349
1789
1734
1488
1417
1069
900
870
833
710
659
465
284

Notes

The spectrum matches that of Adler and Kerr (1963) and is discussed in Farmer (1974).
The spectrum is easily distinguished from that of the higher hydrate **gaylussite.**

References:

1. Huang C.K & Kerr P.F. (1960)
    Infrared study of the carbonate minerals.
    American Mineralogist, **45**, pp. 311-24.

2. Pratt J.H. (1896)
    On northupite; pirssonite, a new mineral; gaylussite and hanksite from Borax Lake, San Bernardino
    Co., California.
    American Journal of Science, **(4)**,2, pp.123 - 35
    Zietschrift für Kristallographie, **27**, pp.416-29
    Yale bicent. Pub., Contr. Mineral., 1901, pp.261-4.

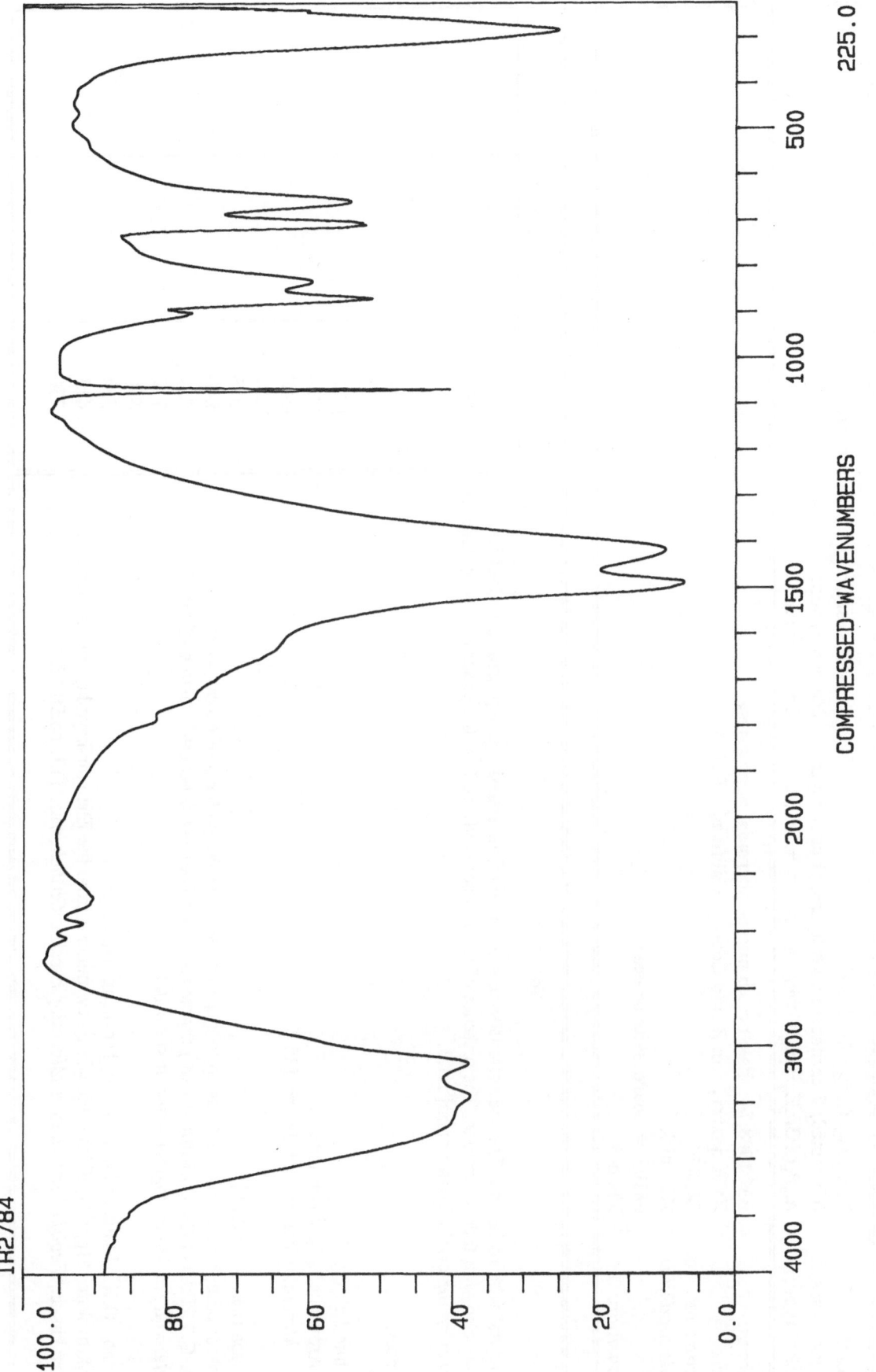

IR2784

% TRANSMITTANCE

100.0   80   60   40   20   0.0

4000   3000   2000   1500   1000   500   225.0

COMPRESSED-WAVENUMBERS

PIRSSONITE

POKROVSKITE

Formula:	$Mg_2(CO_3)(OH)_2 \cdot 0\cdot5\ H_2O$	Crystal system:	Monoclinic
Chemical class:	(Hydrated) ? carbonate with hydroxyl or halogen	Mineral group:	Rosasite
Chemical type:	$A_mB_n(XO_3)_pZ_q \cdot xH_2O$ with (m+n) : p = 2:1	Space group:	$P2_1/a$

Specimen:	BM 1988,74 Pale brown micro-spheroidal aggregates.
Source:	KCA quarry, San Benito County, California, U.S.A.
Spectrum ref. no.:	IR2855
Sample medium:	KBr disk
XRD:	6610F = pokrovskite or near
Composition:	Mg only

Peak Table cm^{-1}

3686	
3575	
3447	
2928	
2291	
1780	
1553	
1426	
1082	
1022	
953	
848	
754	
712	
660	
530	
413	
336	

Notes

Isostructural with malachite. The spectrum is similar to those of the **rosasite** group, and to **malachite**. Compare also with that of **artinite**. The presence of H_2O is not confirmed by the spectrum which matches that given in the original description, ref.3.

References:

1. White J.S. (1987)
   Pokrovskite, a common mineral.
   The Mineralogical Record, **18**, pp.135-6.

2. Fitzpatrick, J.J. (1986)
   Pokrovskite; its possible relationship to mcguinnessite and the problem of excess water.
   In: C.T. Prewitt (Ed) *Abstracts and programme of the Fourteenth general meeting of the International Mineralogical Association*, p.101.

3. Ivanov O.K., Malinovskii Yu.A. & Mozherin Yu.V. (1984)
   Pokrovskite, $Mg_2(CO_3)(OH)_2$: $0\cdot5H_2O$, a new mineral from the Zlatogorskaya layered intrusive, Kazakhstan. *Zapiski Vsesoyunztyi Mineralogicheskoe Obshchestva*, 113, pp.90-95.

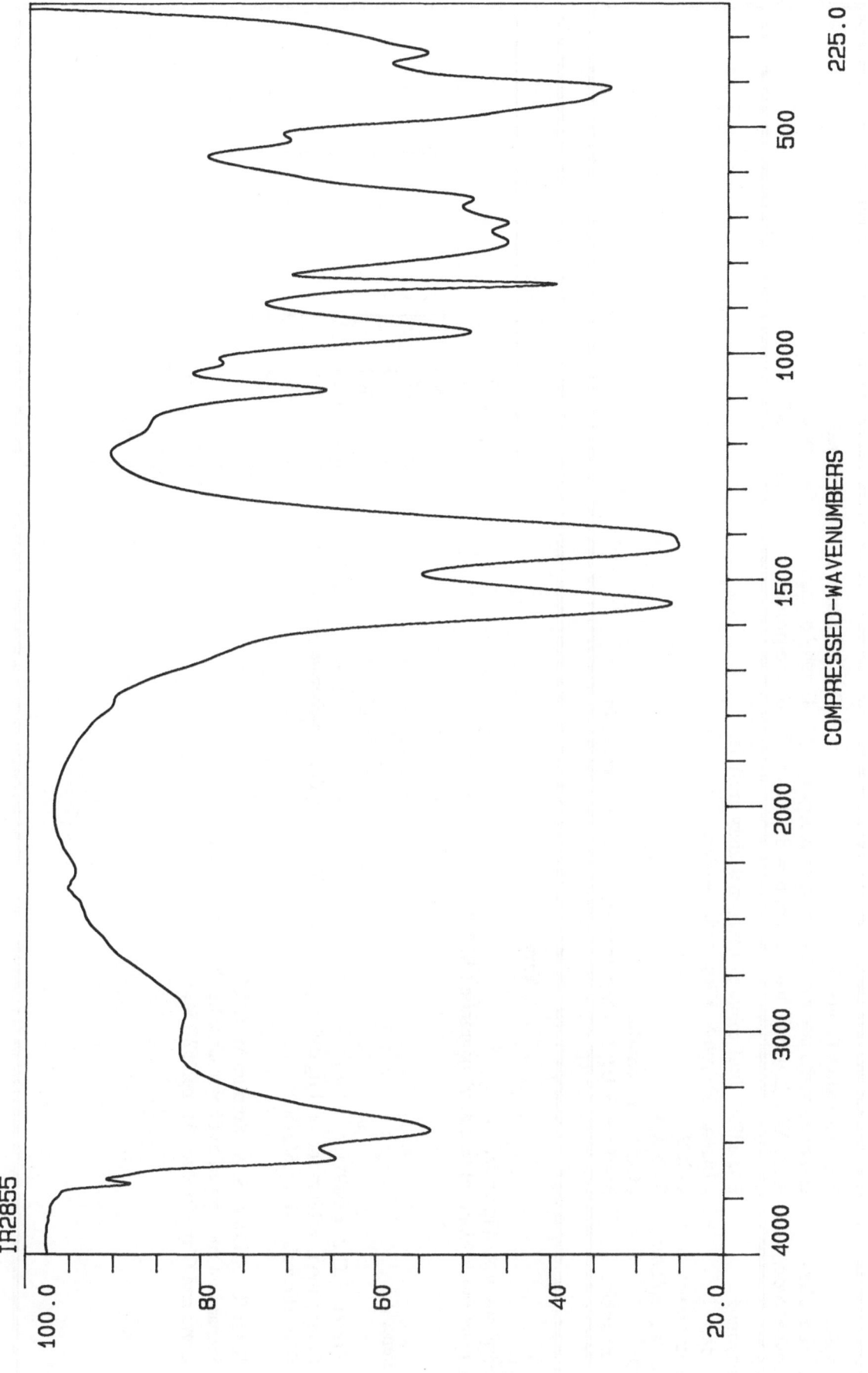

IR2855

% TRANSMITTANCE

COMPRESSED-WAVENUMBERS

POKROVSKITE

PYROAURITE

Crystal system:	**Trigonal**
Mineral group:	**Sjögrenite**
Space group:	**R3̄m or R3m**

Formula:	**$Mg_6Fe_2(CO_3)(OH)_{16} \cdot 4H_2O$**
Chemical class:	**Hydrated carbonate with hydroxyl or halogen**
Chemical type:	**$A_mB_n(XO_3)_pZ_q \cdot nH_2O$** with $(m+n):p = 8:1$

Specimen:	**BM 83815 Buff coloured hexagonal platy crystals.**
Source:	**Långban, Filipstad, Värmland, Sweden.**
Spectrum ref. no.:	**IR2819**
Sample medium:	**KBr disk**
XRD:	**6158F = pyroaurite**
Composition:	**Mg:Fe ≈ 3:1:1 ? with minor Si, Al, Ca and trace Mn**

Peak Table cm^{-1}

3467
2424
1632
1588
1384
1364
1166
1086
685
588
427
377
291

Notes

Dimorphous with sjögrenite.
The spectrum is identical to that of sjögrenite (IR2818).

References:

1. Hansen H.C.B. (1989)
Composition, stabilization, and light absorption of Fe(II)Fe(III) hydroxy carbonate (green rust).
Clay Minerals, **24**, pp.663-669.

2. Hashi K., Kikkawa S. & Koizumi M. (1983)
Preparation and properties of pyroaurite like hydroxy minerals.
Clays and Clay Minerals, **31**, pp.152-154.

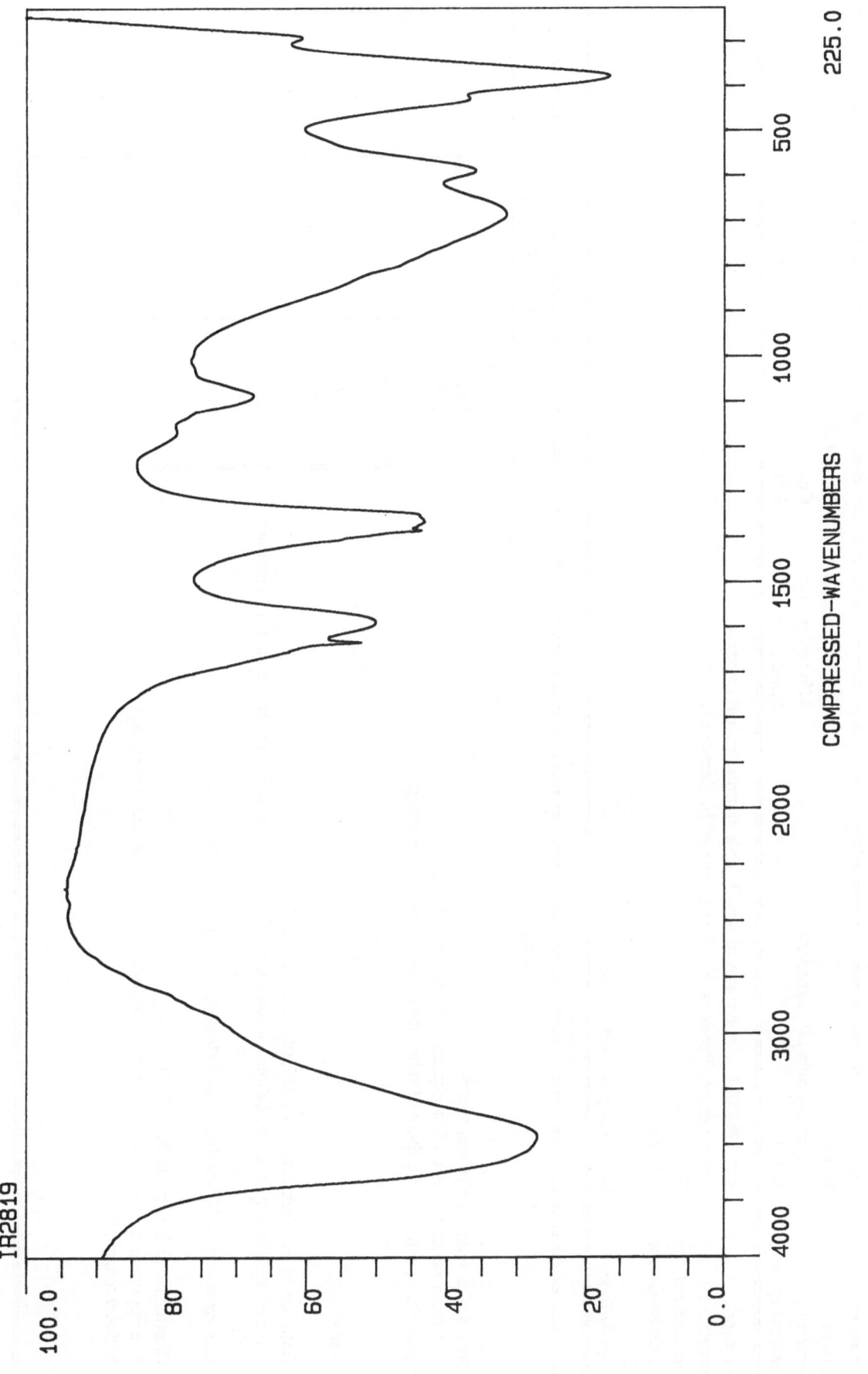

IR2819

% TRANSMITTANCE

100.0

80

60

40

20

0.0

4000

3000

2000

1500

1000

500

225.0

COMPRESSED-WAVENUMBERS

PYROAURITE

RHODOCHROSITE

Crystal system:	Trigonal
Mineral group:	Calcite
Space group:	R3̄c

Formula:	MnCO$_3$
Chemical class:	Anhydrous normal carbonate
Chemical type:	A(XO$_3$)
Specimen:	BM 1984,881 Aggregates of small pink rhombohedral crystals.
Source:	Geevor mine, Pendeen, St. Just in Penwith, Cornwall, U.K.
Spectrum ref. no.:	IR2651
Sample medium:	KBr disk
XRD:	
Composition:	Mn:Ca:Mg ≈ 1·0:0·1:0·03

Peak Table cm⁻¹

[3429]
2924
2850
2580
2486
2128
1801
1420
1085
866
837
726
516
309

Notes

Forms a series with **calcite** and **siderite**.
The spectrum matches those of specimens from other localities.
Compare the spectrum with those of other members of the calcite group.

References:

1. Bottcher M.E., Gehlken P.-L. & Usdowski E. (1992)
   Infrared spectroscopic investigations of the calcite-rhodochrosite and parts of the calcite-magnesite series.
   Contributions to Mineralogy and Petrology, **109**, pp.304-306.

2. Chester R. & Elderfield H. (1967)
   The application of infra-red absorption spectroscopy to carbonate mineralogy.
   Sedimentology, **9**, pp.5307-9.

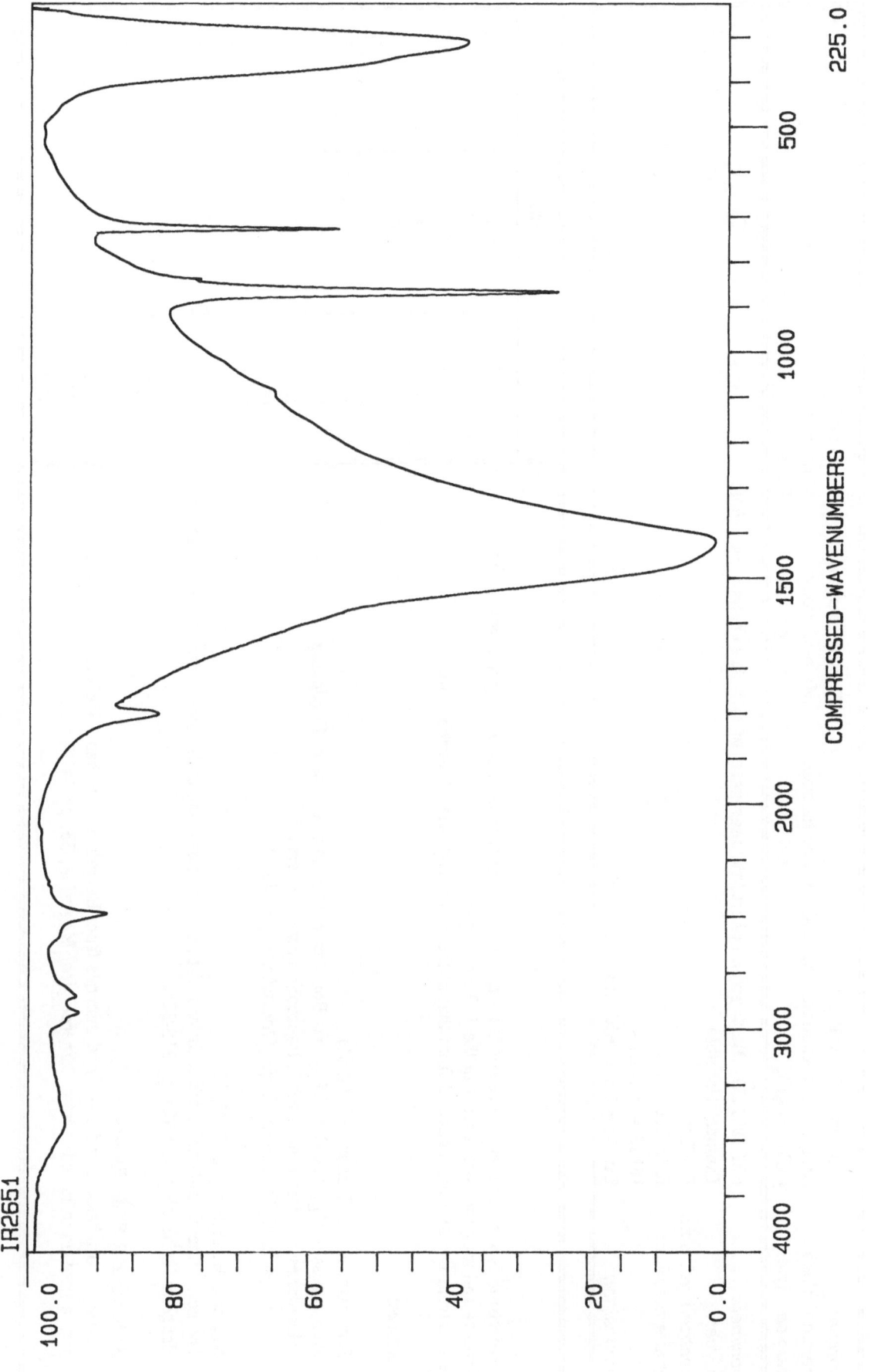

% TRANSMITTANCE

IR2651

COMPRESSED-WAVENUMBERS

RHODOCHROSITE

225.0
500
1000
1500
2000
3000
4000

100.0
80
60
40
20
0.0

ROSASITE

Formula:	$(Cu,Zn)_2(CO_3)(OH)_2$
Chemical class:	Anhydrous carbonate with hydroxyl or halogen
Chemical type:	$(AB)_2(XO_3)Z_q$

Crystal system:	Monoclinic
Mineral group:	Malachite
Space group:	$P2_1/a$

Specimen:	BM 1972,33 Dark green spherulitic aggregates of lath-like crystals on calcite.
Source:	Tsumeb, Namibia.
Spectrum ref. no.:	IR2738
Sample medium:	KBr disk
XRD:	16163 = rosasite
Composition:	Cu:Zn ≈ 1·6:1 with trace Mg & Si

Peak Table cm⁻¹

3494	555
3427	459
3245	409
2928	330
2545	277
2403	
2068	
1780	
1515	
1419	
1384	
1165	
1100	
1049	
854	
828	
739	
706	
671	

Notes

The spectrum is very similar to those of other members of the rosasite group i.e. **glaukosphaerite,
kolwezite** and **mcguinnessite**, except in the 700 cm⁻¹ region.

The spectrum is easily distinguished from that of the chemically related **aurichalcite.**

References:

1. Schmetzer K. & Tremmel G. (1981)
    Mcguinnessit $(Mg,Cu)_2 CO_3 (OH)_2$ aus Bou Azzer, Marokko; ein neuer Fundpunkt
    (Mcguinnessite from Bou Azzer, Morocco; a new discovery).
    Neues Jahrbuch für Mineralalogie, Monatshefte, pp.443-51.

2. Nickel E.H. & Berry L.G. (1981)
    The new mineral nullaginite and additional data on the related minerals rosasite and glaukosphaerite.
    Canadian Mineralogist, **19**(2), pp.315-324.

3. Braithwaite R.S. & Ryback G. (1963)
    Rosasite, aurichalcite, and associated minerals from Heights of Abraham, Matlock Bath, Derbyshire,
    with a note on infra-red spectra. *Mineralogical Magazine*, **33**, pp.441-449.

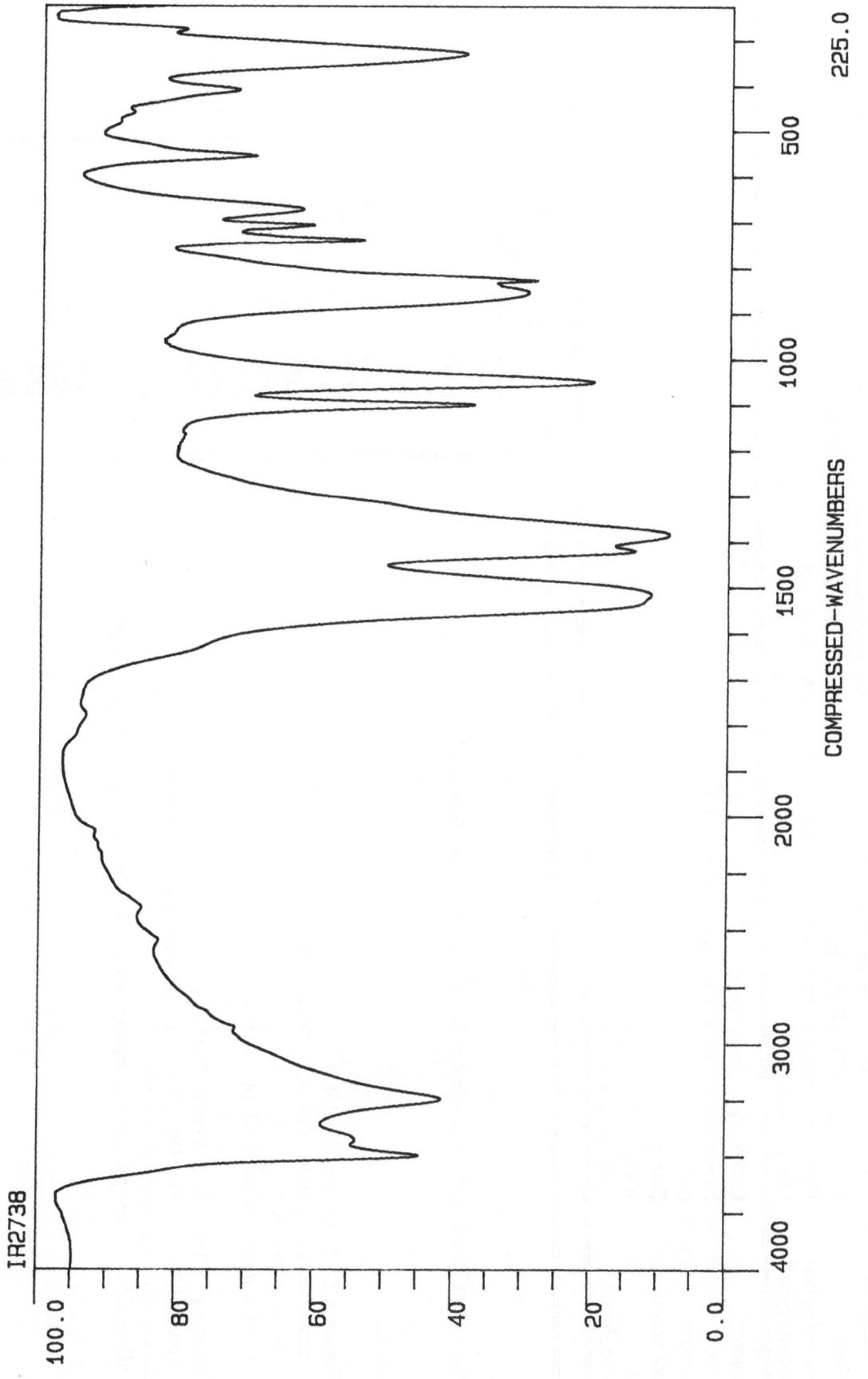

IR2738

% TRANSMITTANCE

COMPRESSED-WAVENUMBERS

ROSASITE

ROUBAULTITE

Formula:	$Cu_2O_2(UO_2)_3(CO_3)_2(OH)_2 \cdot 4H_2O$
Chemical class:	Hydrated carbonate with hydroxyl or halogen
Chemical type:	Miscellaneous
Specimen:	C3663 Pale green fibrous.
Source:	Kamoto-Olivera-Virgule mine, Shaba, Zaïre.
Spectrum ref. no.:	IR2915
Sample medium:	KBr disk
XRD:	4568
Composition:	Cu:U = 2:3·6

Crystal system:	Triclinic
Mineral group:	
Space group:	P$\bar{1}$

Notes

Specimen from Institut Royal des Sciences Naturelle de Belgique, Brussels.

References:

1. Ginderow D. & Cesbron F. (1985)
    Structure de la roubaultite, $Cu_2(UO_2)_3(CO_3)_2O_2(OH)_2 \cdot 4H_2O$
    (Structure of roubaultite).
    Acta Crystallographica, **41**(C), pp.654-657.

2. Cesbron F., Pierrot R. & Verbeek T. (1970)
    La roubaultite $Cu_2(UO_2)_3(OH)_{10} \cdot 5H_2O$ une nouvelle espèce minérale.
    (Roubaultite a new mineral species).
    Bulletin de la Société française de Minéralogie et de Cristallographie, **93**, pp.550-554.

Peak Table cm^{-1}

3522	791
3397	755
3306	733
3196	708
2646	524
2615	469
2516	421
2339	318
2025	279
1859	246?
1836	
1733	
1638	
1504	
1399	
1148	
1015	
894	
824	

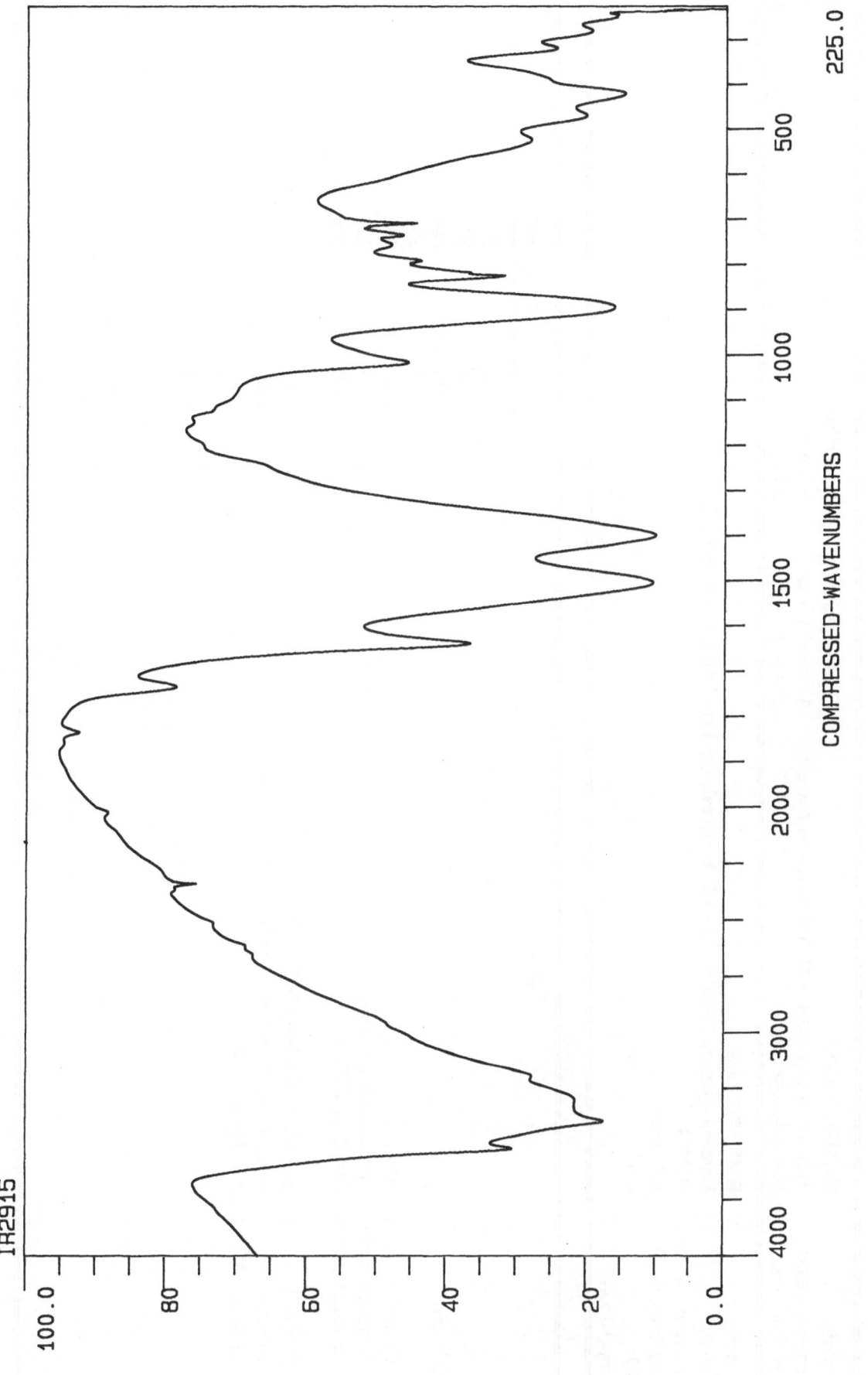

IR2915

% TRANSMITTANCE

100.0　80　60　40　20　0.0

4000　3000　2000　1500　1000　500　225.0

COMPRESSED-WAVENUMBERS

ROUBAULTITE

SABINAITE

Formula:	Na$_4$TiZr$_2$O$_4$(CO$_3$)$_4$
Chemical class:	Anhydrous carbonate with hydroxyl and halogen
Chemical type:	Miscellaneous

Crystal system:	Monoclinic
Mineral group:	Sabinaite
Space group:	C2/c

Specimen:	RMS 1982.25.6.
Source:	Francon Quarry, Mont St. Michel, Montreal, Quebec, Canada. (Type locality)
Spectrum ref. no.:	IR2914
Sample medium:	KBr disk
XRD:	4116
Composition:	

Peak Table cm^{-1}

[3390]	747
3120	691
2926	646
2858	474
2636	391
2369	334
2339	289
1826	
1773	
1681	
1653	
1567	
1324	
1086	
1066	
856	
830	
811?	
771	

Notes

References:

1. Chao G.Y. & Gu J. (1985)
   Sabinaite; a new occurrence and new data.
   Canadian Mineralogist, **23**, pp.17-19.

2. Jambor J.L., Sturman B.D. & Weatherly G.C. (1980)
   Sabinaite, a new anhydrous zirconium bearing carbonate mineral from Montreal Island, Quebec.
   Canadian Mineralogist, **18**, pp.25-29.

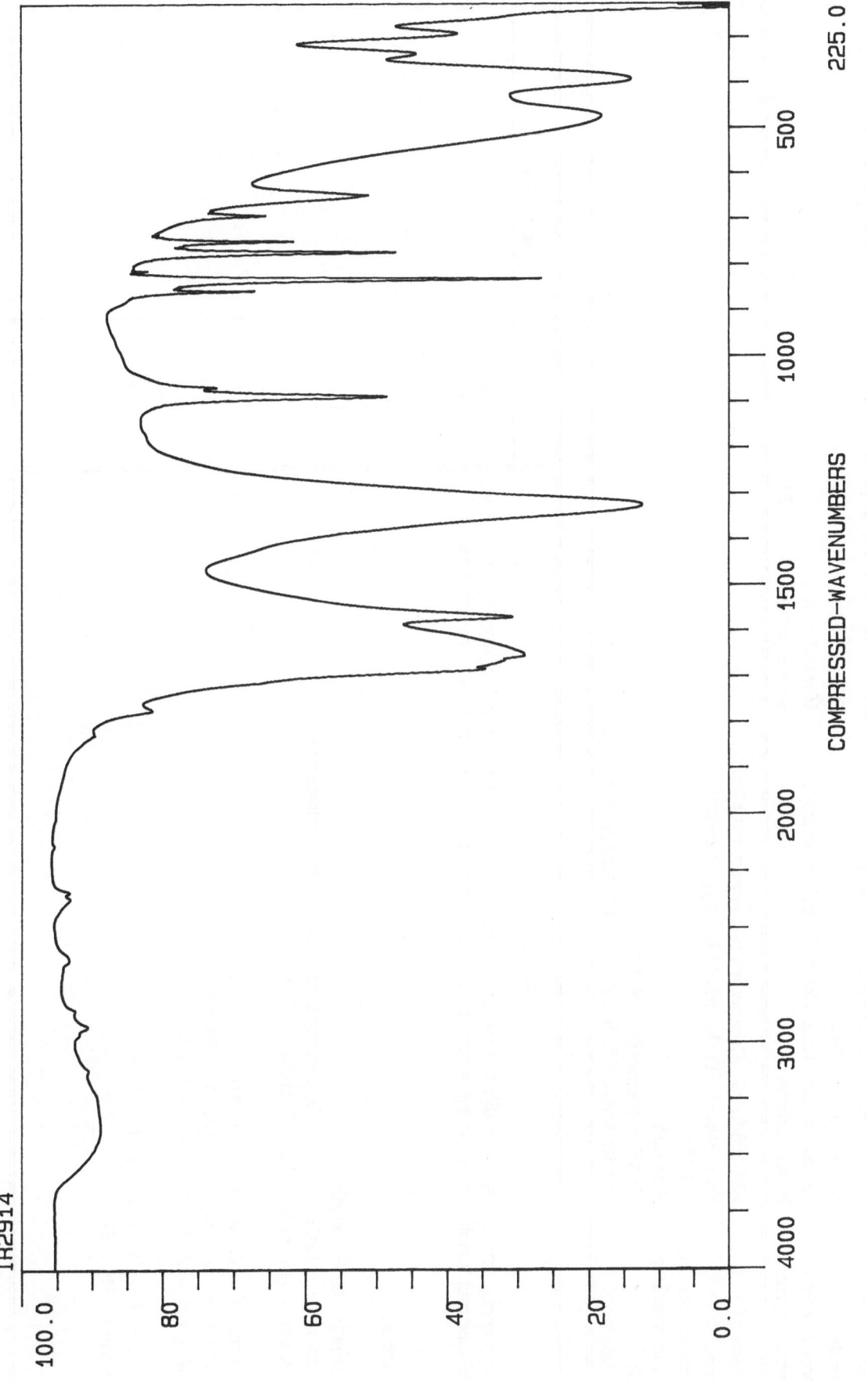

IR2914

% TRANSMITTANCE

100.0

80

60

40

20

0.0

4000   3000   2000   1500   1000   500   225.0

COMPRESSED-WAVENUMBERS

SABINAITE

SCARBROITE

Formula:	Al$_5$(OH)$_{13}$(CO$_3$)·5H$_2$O
Chemical class:	**Hydrated carbonate with hydroxyl or hydrogen**
Chemical type:	**Miscellaneous**
Crystal system:	Triclinic
Mineral group:	
Space group:	**P1**

Specimen:	**BM 1984,898 Creamy white compact with halite.**
Source:	Scarborough, Yorkshire, U.K. (type locality)
Spectrum ref. no.:	IR2798
Sample medium:	KBr disk
XRD:	4208F = scarbroite + halite
Composition:	Al with minor Cl, Na, Si, S and trace Ca, Mg

Peak Table cm^{-1}

3614
3413
2123
1633
1462
1424
1100
1019
980
744
623
552
509
392
356
319

Notes

The specimen contains impurities due to exposure to sea water. The major one, halite, does not contribute to the spectrum which is close to that given in ref. 2, but some relative peak intensities are different.

References:

1. Brindley G.W. (1980)
    Scarbroite, Al$_5$(OH)$_{13}$CO$_3$·5H$_2$O, compared with gibbsite and hydrotalcite.
    Mineralalogical Magazine, **43**, pp.615-618.

2. Duffin W.J. & Goodyear J. (1960)
    A thermal and x-ray investigation of scarbroite.
    Mineralogical Magazine, **32**, pp.353-362.

3. Duffin W.J. & Goodyear J. (1957)
    Nature, **180**, p.977.

IR2798

% TRANSMITTANCE

COMPRESSED-WAVENUMBERS

SCARBROITE

SCHRÖCKINGERITE

Formula:	$NaCa_3(UO_2)(CO_3)_3(SO_4)F \cdot 10H_2O$
Chemical class:	Compound carbonate
Chemical type:	Miscellaneous
Crystal system:	Triclinic
Mineral group:	
Space group:	P1

Specimen:	RMS 1978.17.98.
Source:	White Canyon mine, Frey Point, San Juan Co., Utah, California, USA.
Spectrum ref. no.:	IR2918
Sample medium:	KBr disk
XRD:	3896
Composition:	

Peak Table cm^{-1}

3598	1082
3468	986
3272	**906**
2964	**843**
2933	822
2656	741
2457	706
2413	684
2363	**612**
2334	**546**
2081	**434**
1825	**286**
1718	**252?**
1643	
1577	
1550	
1370	
1187	
1098	

Notes

References:

1. Urbanec Z. & Čejka J. (1979)
   Infrared spectra of liebigite, andersonite, voglite, and schröckingerite.
   Collection of Czechoslovak Chemical Communications, **44**, pp.10-23.

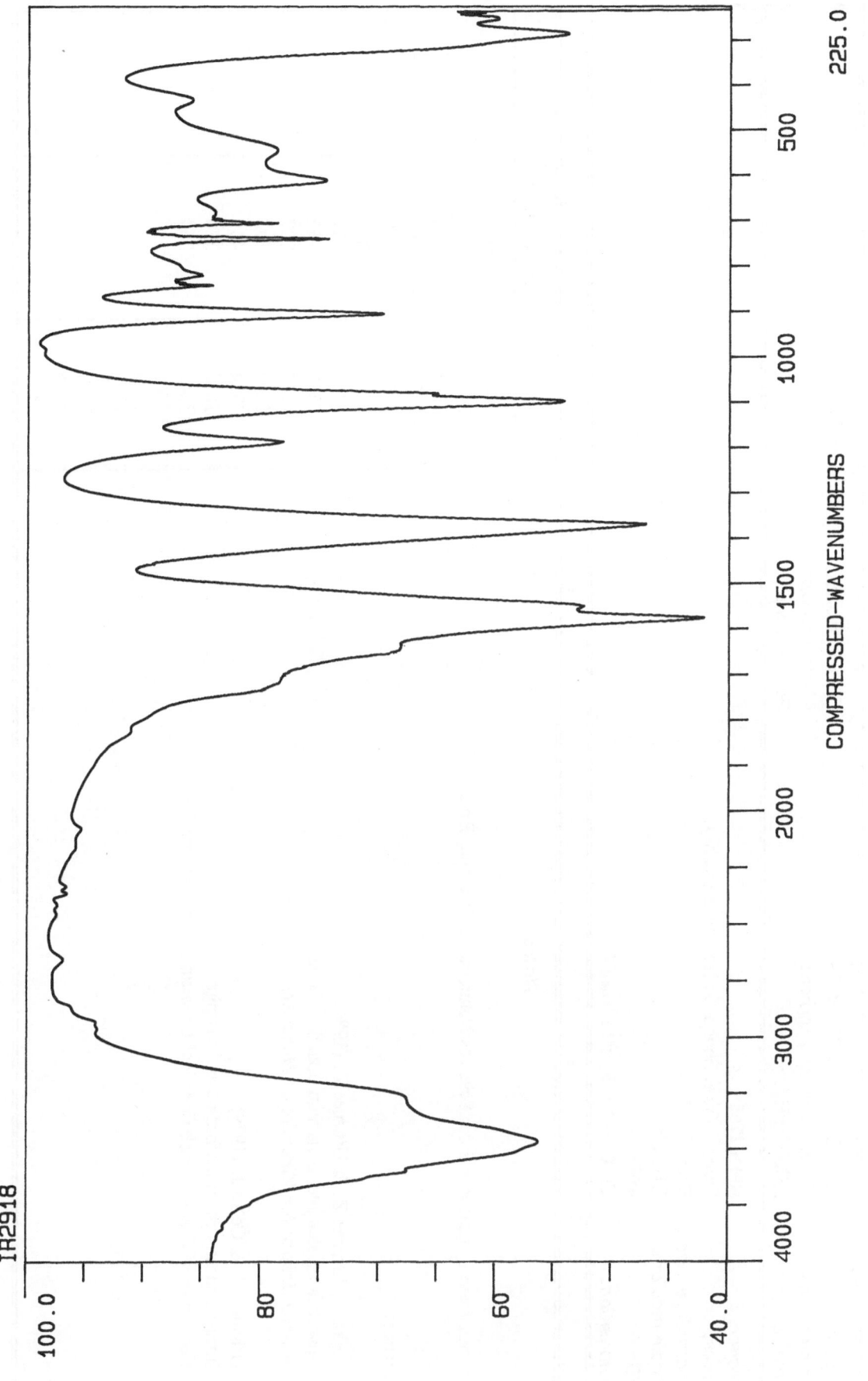

IR2918

% TRANSMITTANCE

100.0

80

60

40.0

4000

3000

2000

1500

1000

500

225.0

COMPRESSED-WAVENUMBERS

SCHRÖCKINGERITE

SHARPITE

Crystal system:	Orthorhombic
Mineral group:	
Space group:	?

Formula:	$Ca(UO_2)_6(CO_3)_5(OH)_4 \cdot 6H_2O$
Chemical class:	Hydrated normal carbonate
Chemical type:	$A(XO_3) \cdot xH_2O$

Specimen:	RMS 2768
Source:	Shinkolobwe, Shaba, Zaïre. (type locality)
Spectrum ref. no.:	IR2943
Sample medium:	KBr disk
XRD:	4564
Composition:	Ca:U ≈ 1:5·5 with trace Fe

Peak Table cm⁻¹

3546	1101
3437	959
3243	916
3001	848
2956	828
2646	814
2499	777
2339	762
2217	707
1866	692
1734	459
1627	374
1543	254
1460	
1449	
1420	
1244	
1194	
1151	

Notes

Specimen from Institut Royal des Sciences Naturelle de Belgique, Brussels.

References:

1. Čejka J., Mrázek Z. & Urbanec Z. (1984)
   New data on sharpite, a calcium uranyl carbonate.
   Neues Jahrbuch für Mineralogie, Monatshefte, pp.109-117.

2. Urbanec Z. & Čejka J. (1979)
   Infrared spectra of rutherfordine and sharpite.
   Collection of Czechoslovak Chemical Communications, **44**, pp.1-9.

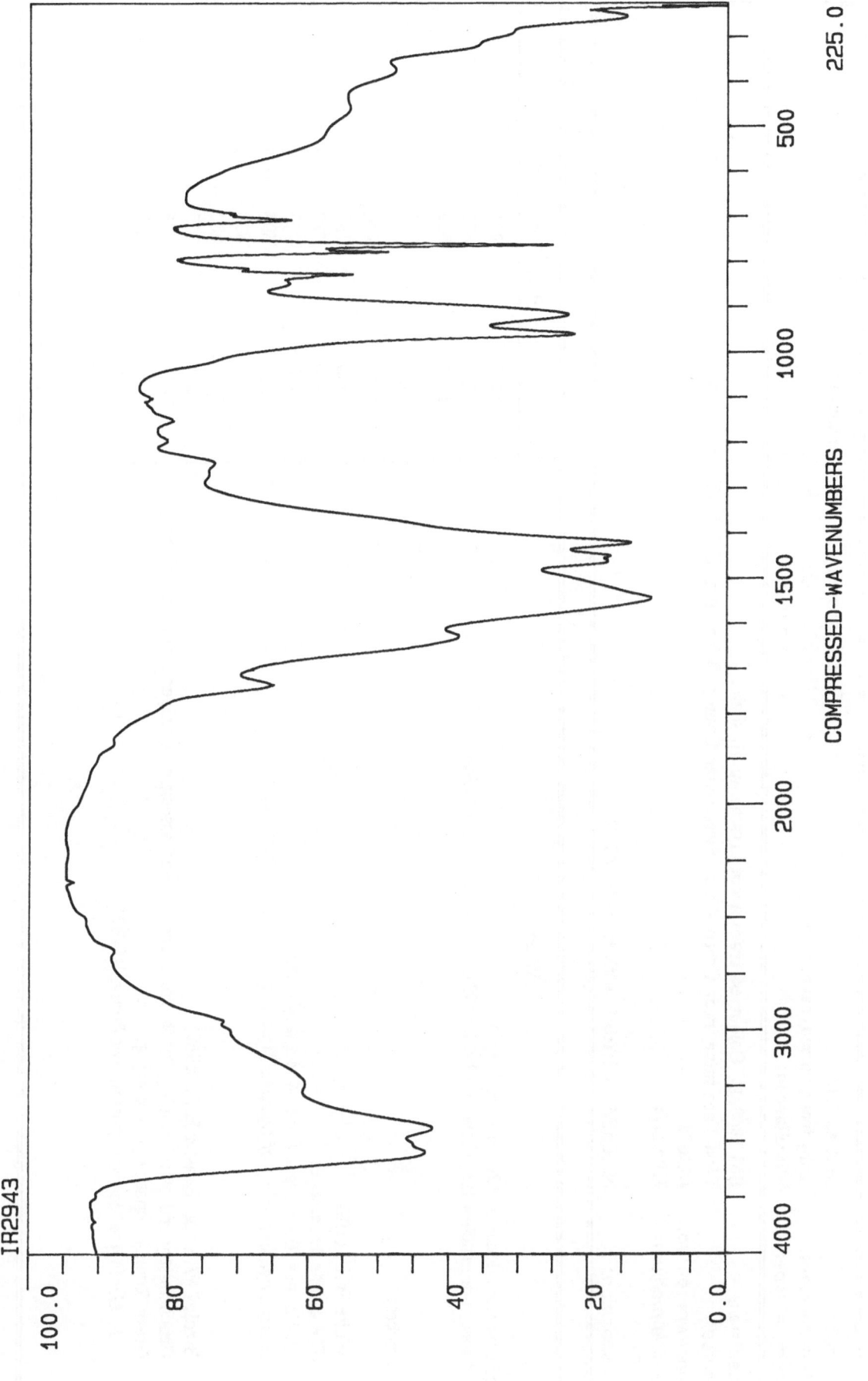

IR2943

% TRANSMITTANCE

COMPRESSED-WAVENUMBERS

SHARPITE

SHORTITE

Formula:	Na$_2$Ca$_2$(CO$_3$)$_3$
Chemical class:	Anhydrous normal carbonate
Chemical type:	Miscellaneous

Crystal system:	Orthorhombic
Mineral group:	Eitelite
Space group:	Amm2

Specimen:	BM 1968,42 Colourless translucent crystals in oil shale.
Source:	West Vaca mine, near Green River, Sweetwater County, Wyoming, U.S.A. (Type locality).
Spectrum ref. no.:	IR2670
Sample medium:	KBr disk
XRD:	
Composition:	Na:Ca:Mg ≈ 1:1:0·1 with trace K, Al, Si

Notes

See expanded detail of 1200-600 cm^{-1} region.
The spectrum matches that shown in Sadtler (93), but has better resolution.

References:

1. White W.B. (1974)
The carbonate minerals.
In: Farmer (Ed.) *The Infrared Spectra of Minerals.*
Mineralogical Society of London, Monograph No. 4, pp.227-284.

2. Bradley W.H. & Eugster H.P. (1969)
Geochemistry and paleolimnology of the trona deposits and associated authigenic minerals of the
Green River Formation of Wyoming.
U.S. Geological Survey Professional Paper, B1-B71.

Peak Table cm^{-1}

[3527]	891
2990	872
2941	866
2905	851
2836	842
2616	826
2548	731
2509	719
2460	710
2386	694
1808	686
1765	682
1521	325
1481	286
1453	257
1409	
1090	
1071	
1048	

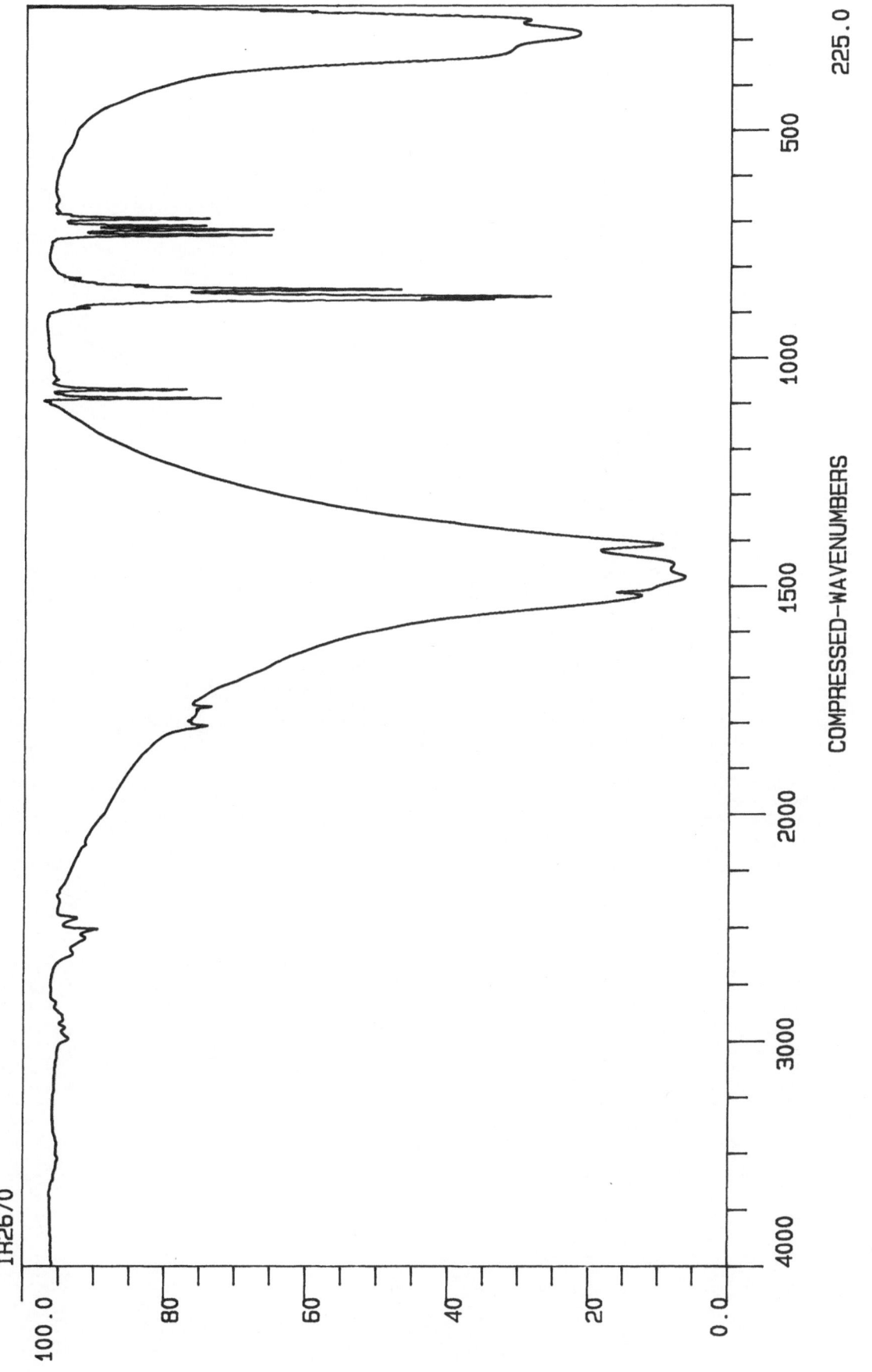

IR2670

% TRANSMITTANCE

COMPRESSED-WAVENUMBERS

SHORTITE

4000   3000   2000   1500   1000   500   225.0

100.0   80   60   40   20   0.0

IR2670

% TRANSMITTANCE

WAVENUMBERS

SHORTITE   [expanded detail]

SIDERITE

Formula:	FeCO$_3$
Chemical class:	Anhydrous normal carbonate
Chemical type:	A(XO$_3$)
Crystal system:	Trigonal
Mineral group:	Calcite
Space group:	R$\bar{3}$c

Specimen:	BM 1929,219 Pale yellow lenticular crystals.
Source:	New Wheal Kitty, St Agnes, Cornwall, U.K.
Spectrum ref. no.:	IR2650
Sample medium:	KBr disk
XRD:	
Composition:	Fe with trace Al, Si

Peak Table cm^{-1}

[3285]
2925
2856
2500
1811
1422
1094
867
739
605
336

Notes

Forms a series with **magnesite** and **rhodochrosite**.
The spectrum matches those of specimens from other localities.
Compare the spectrum with those of other members of the **calcite** group.

References:

1. Dubrawski J.V., Channon A.L. & Warne S.S.J. (1989)
   Examination of the siderite magnesite mineral series by Fourier transform infrared spectroscopy.
   American Mineralogist, **74**, pp.187-190.

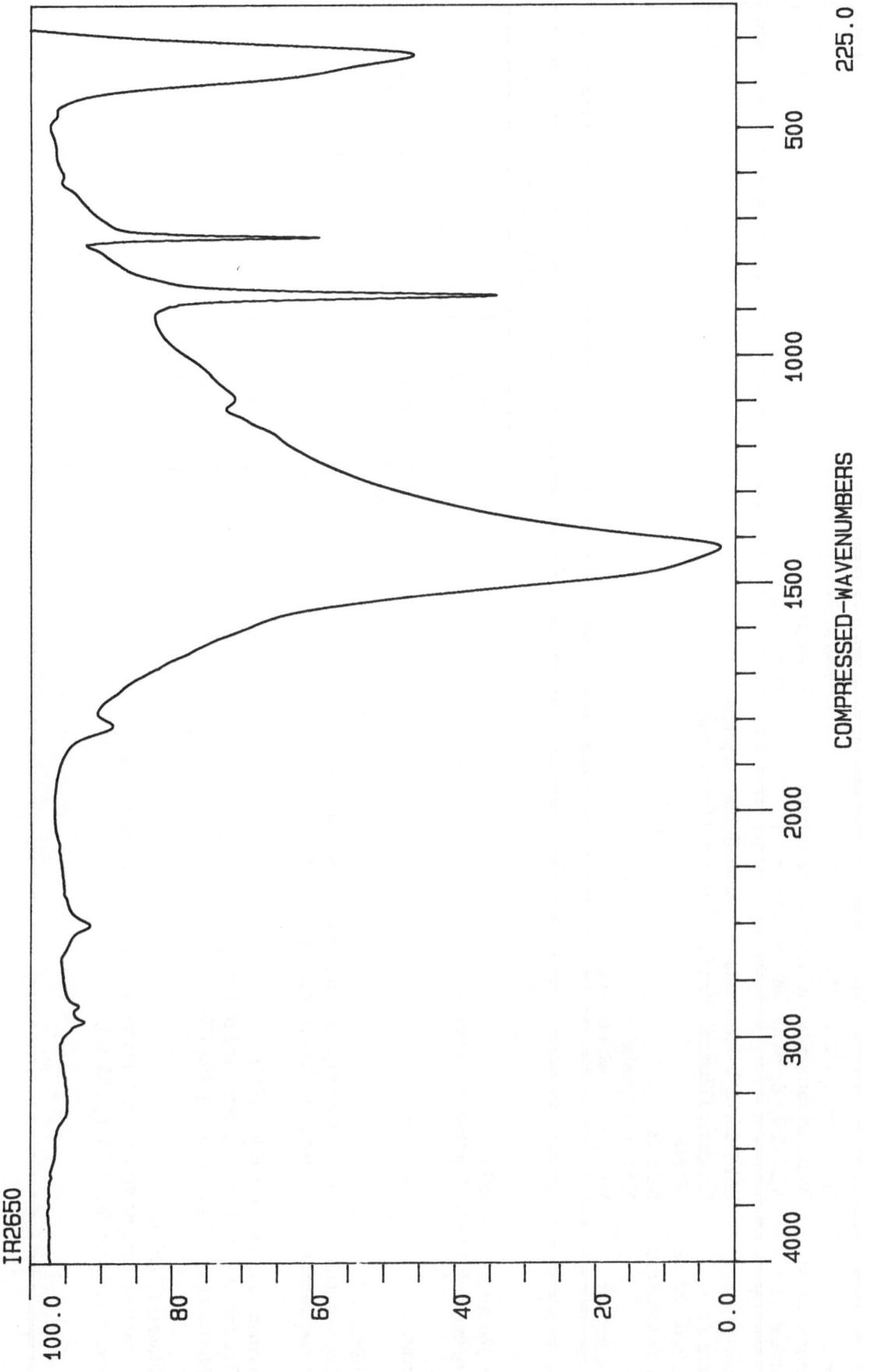

IR2650

% TRANSMITTANCE

COMPRESSED-WAVENUMBERS

SIDERITE

SJÖGRENITE

Formula:	$Mg_6Fe_2^{3+}(CO_3)(OH)_{16}\cdot 4H_2O$
Chemical class:	Hydrated carbonate with hydroxyl or halogen
Chemical type:	$A_mB_n(XO_3)_pZ_q\cdot xH_2O$ with $(m+n):p = 8:1$
Crystal system:	Hexagonal
Mineral group:	
Space group:	$P6_3/mmc$

Specimen:	BM 1926,1222 Straw coloured, isolated, platy crystals.
Source:	Långban, Filipstad, Värmland, Sweden. (Type locality).
Spectrum ref. no.:	IR2818
Sample medium:	KBr disk
XRD:	6161 = sjögrenite
Composition:	Mg:Fe ≈ 3:1 with trace Si

Peak Table cm⁻¹

3469
2925
2434
2366
1632
1589
1384
1366
1167
1086
1020
993
683
584
428
377
292

Notes

Dimorphous with pyroaurite.
The spectrum is identical to that of pyroaurite.

References:

1. Allmann R. (1969)
    Supplemental information on the structures of pyroaurite and sjögrenite.
    Neues Jahrbuch für Mineralogie, Monatshefte, pp.552-558.

2. Ingram L. & Taylor H.F.W. (1967)
    The crystal structures of sjögrenite and pyroaurite.
    Mineralogical Magazine, **36**, pp.465-479.

3. Frondel C. (1941)
    Constitution and polymorphism of the pyroaurite and sjögrenite groups.
    American Mineralogist, **26**, pp.295-315.

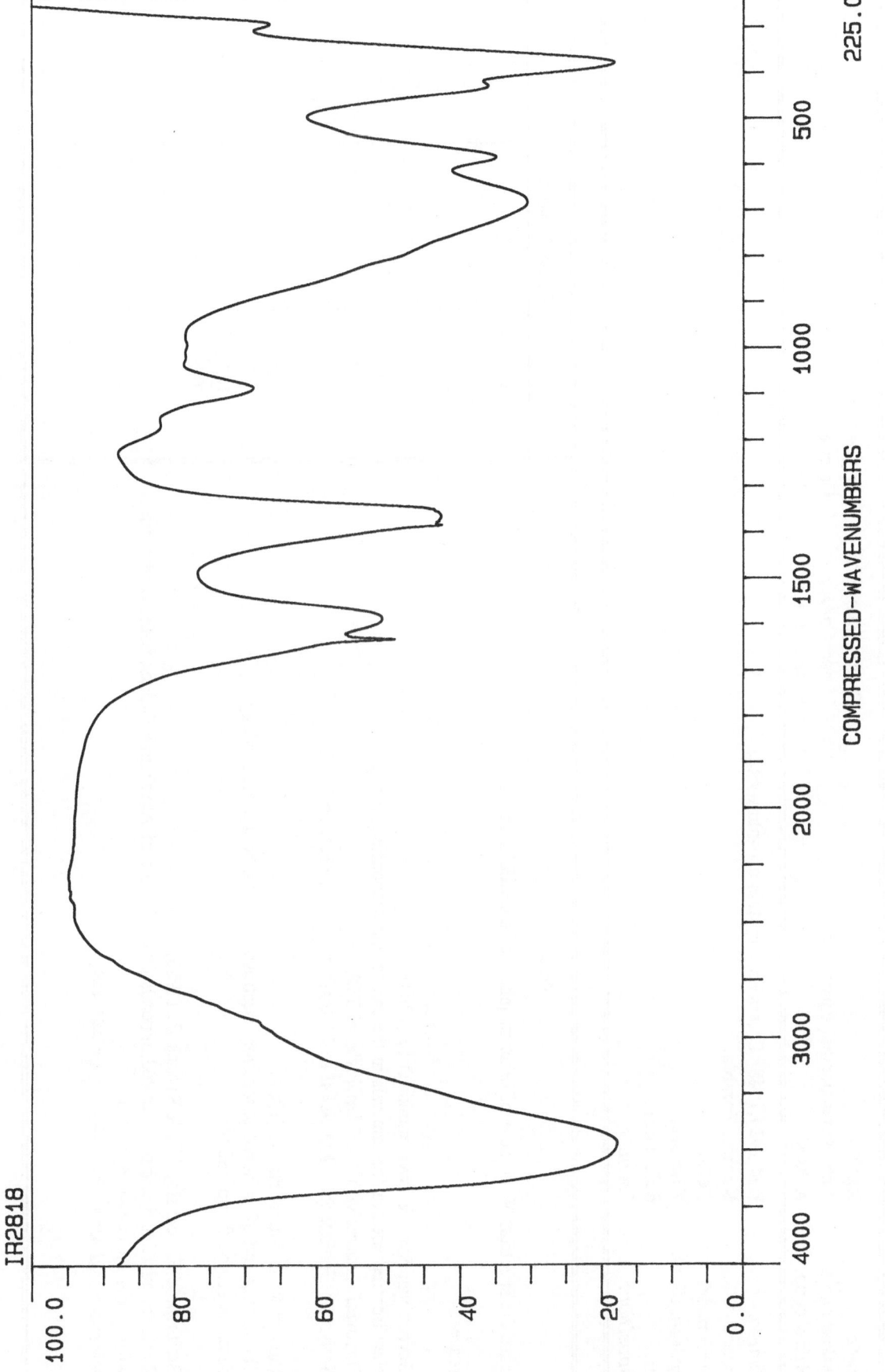

IR2818

% TRANSMITTANCE

COMPRESSED-WAVENUMBERS

SJOGRENITE

SMITHSONITE

Crystal system:	**Trigonal**
Mineral group:	**Calcite**
Space group:	**R3̄c**

Formula:	**ZnCO₃**	
Chemical class:	**Anhydrous normal carbonate**	
Chemical type:	**A(XO₃)**	
Specimen:	**BM 1929,1648**	**Colourless crystals on sphalerite.**
Source:	**Kabwe, Zambia.**	
Spectrum ref. no.:	**IR2649**	
Sample medium:	**KBr disk**	
XRD:	**8273 (std)**	
Composition:	**Zn only**	

Formula: **ZnCO₃**

Peak Table cm⁻¹

[3419]
2924
2850
2493
1816
1605
1427
1170
1096
870
841
745
308

Notes

Compare the spectrum with those of other members of the **calcite group**.

References:

1. Gevork'yan S.V. & Povarennikh O.S. (1983)
   New infrared spectra for minerals in the calcite and aragonite groups.
   Dopovidi Akademiyi Nauk Ukrayins'koyi RSR,
   Seriya B: Geologichni, Khimichni ta Biologichni Nauki, (11), pp.8-12.

2. Chester R. & Elderfield H. (1967)
   The application of infra-red absorption spectroscopy to carbonate mineralogy.
   Sedimentology, **9**, pp.5307-9.

3. Braithwaite R., Stanley W. & Ryback G. (1963)
   Rosasite, aurichalcite and associated minerals from Heights of Abraham, Matlock Bath, Derbyshire,
   with a note on infra-red spectra.
   Mineralogical Magazine, **33**(261), pp.441-449.

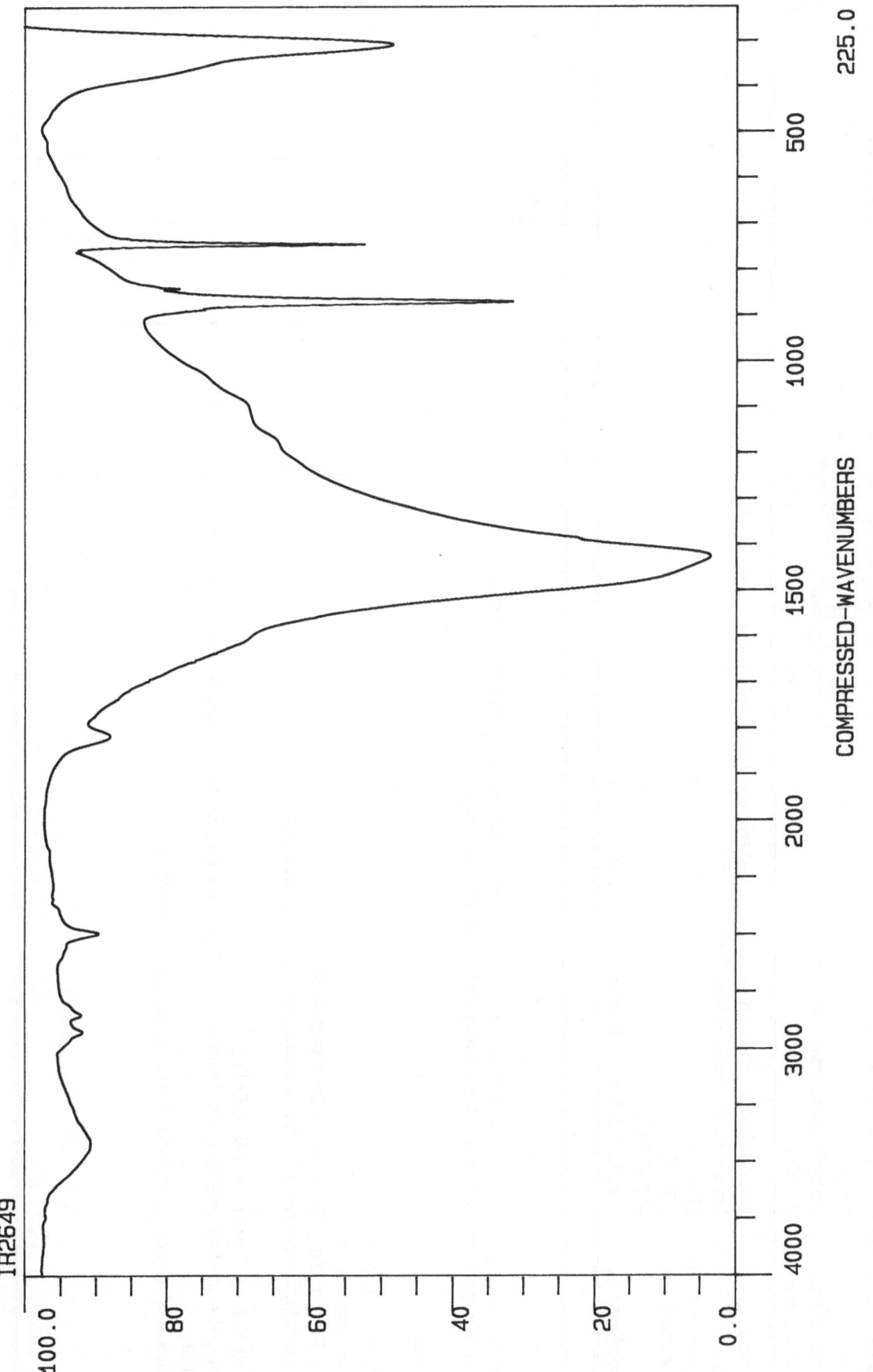

IR2649

% TRANSMITTANCE

COMPRESSED-WAVENUMBERS

SMITHSONITE

SPHAEROCOBALTITE

Formula:	$CoCO_3$
Chemical class:	Anhydrous normal carbonate
Chemical type:	$A(XO_3)$

Crystal system:	**Trigonal**
Mineral group:	**Calcite**
Space group:	**R3̄c**

Specimen:	BM 1967,287 Small dark pink, flattened rhombs on dolomite matrix.
Source:	**Musonoi, Kolwezi, Shaba, Zaïre.**
Spectrum ref. no.:	IR2652
Sample medium:	KBr disk
XRD:	6259F (std)
Composition:	Co:Mg ≈ 1:0·4 with trace Ca,Fe,Mn,Sr

Peak Table cm⁻¹

Peak Table cm^{-1}

3227
2927
2861
2509
1819
1427
1120
1092
875
747
516
372
245?

Notes

Specimens with high cobalt content are rare and many 'sphaerocobaltites' are cobaltian dolomite.
Compare the spectrum with those of other members of the **calcite** group.

References:

1. White W.B. (1974)
    The carbonate minerals.
    In: Farmer (Ed.) *The Infrared Spectra of Minerals.*
    Mineralogical Society of London, Monograph No.4, pp.227-284.

2. Weir C.E. & Lippincott E.R. (1961)
    Infrared studies of aragonite, calcite and vaterite type structures in the borates, carbonates and nitrates.
    Journal of Research, National Bureau of Standards, **65**(A), pp.173-183

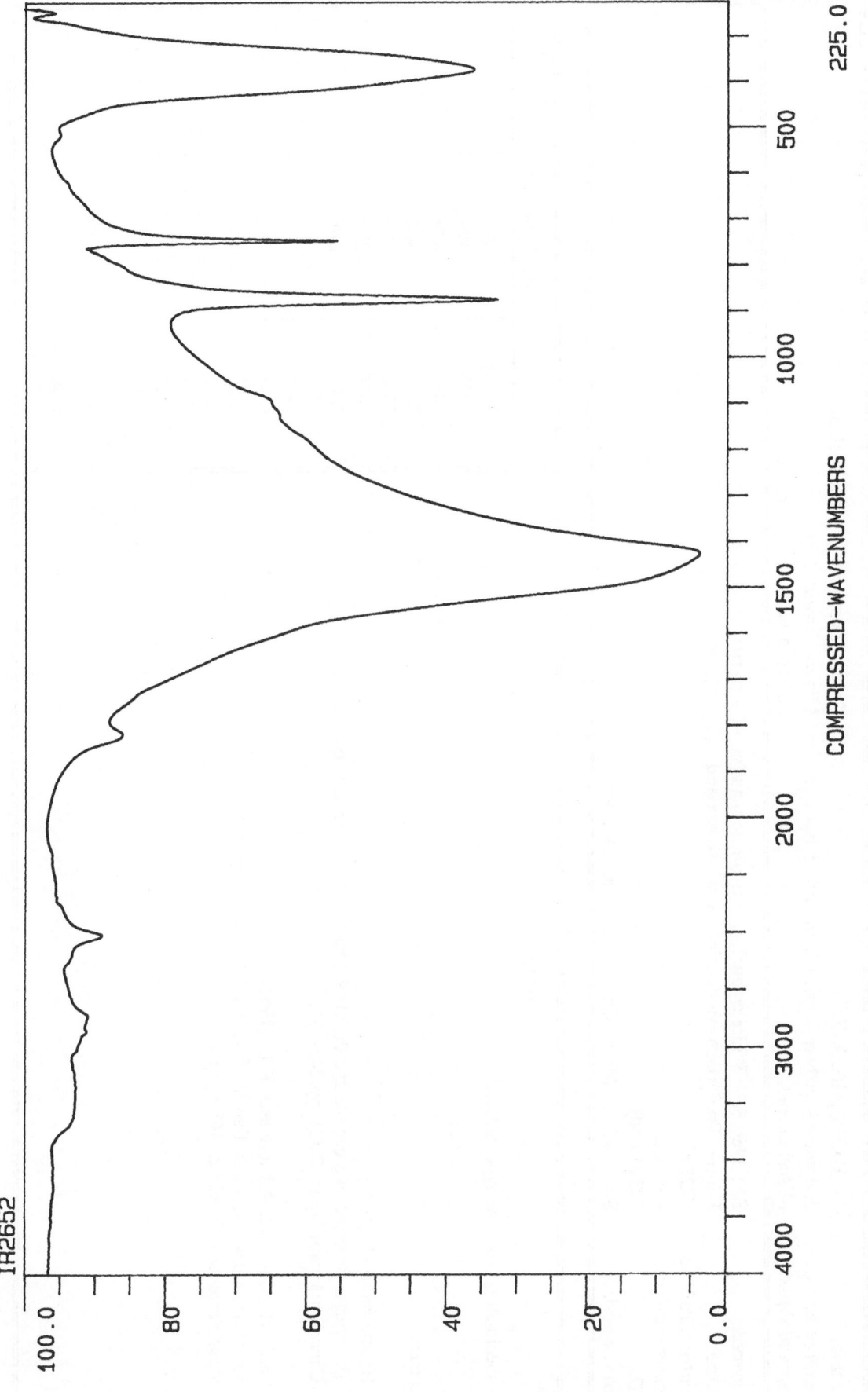

IR2652

% TRANSMITTANCE

COMPRESSED-WAVENUMBERS

SPHAEROCOBALTITE

STENONITE

Formula:	(Sr,Ba,Na)$_2$Al(CO$_3$)F$_5$
Chemical class:	Anhydrous carbonate with hydroxyl or halogen
Chemical type:	Miscellaneous
Crystal system:	Monoclinic
Mineral group:	
Space group:	P2$_1$/m

Specimen:	BM 1966,536 White/colourless crystalline massive with pyrite, sphalerite etc.
Source:	Ivigtut, Frederikshaab district, South Greenland. (Type locality).
Spectrum ref. no.:	IR2815
Sample medium:	KBr disk
XRD:	12313 (std)
Composition:	Sr:Al ≈ 2:1 Minor Ca,Si, trace only Ba, Na

Peak Table cm^{-1}

[3341]	551
2935	496
2857	429
2572	404
2539	339
1846	299
1807	246?
1486	
1432	
1098	
1027	
870	
843	
799	
751	
710	
702	
606	

Notes

The specimen is from the type material.

References:

1. Hawthorne F.C. (1984)
   The crystal structure of stenonite and the classification of the aluminofluoride minerals.
   Canadian Mineralogist, **22**(2), pp.245-251.

2. Pauly H., Dano M. & Mortensen E.L. (1962)
   Stenonite, a new carbonate fluoride from Ivigtut, South Greenland.
   Meddelelser om Grønland, **169**(9), p.24.

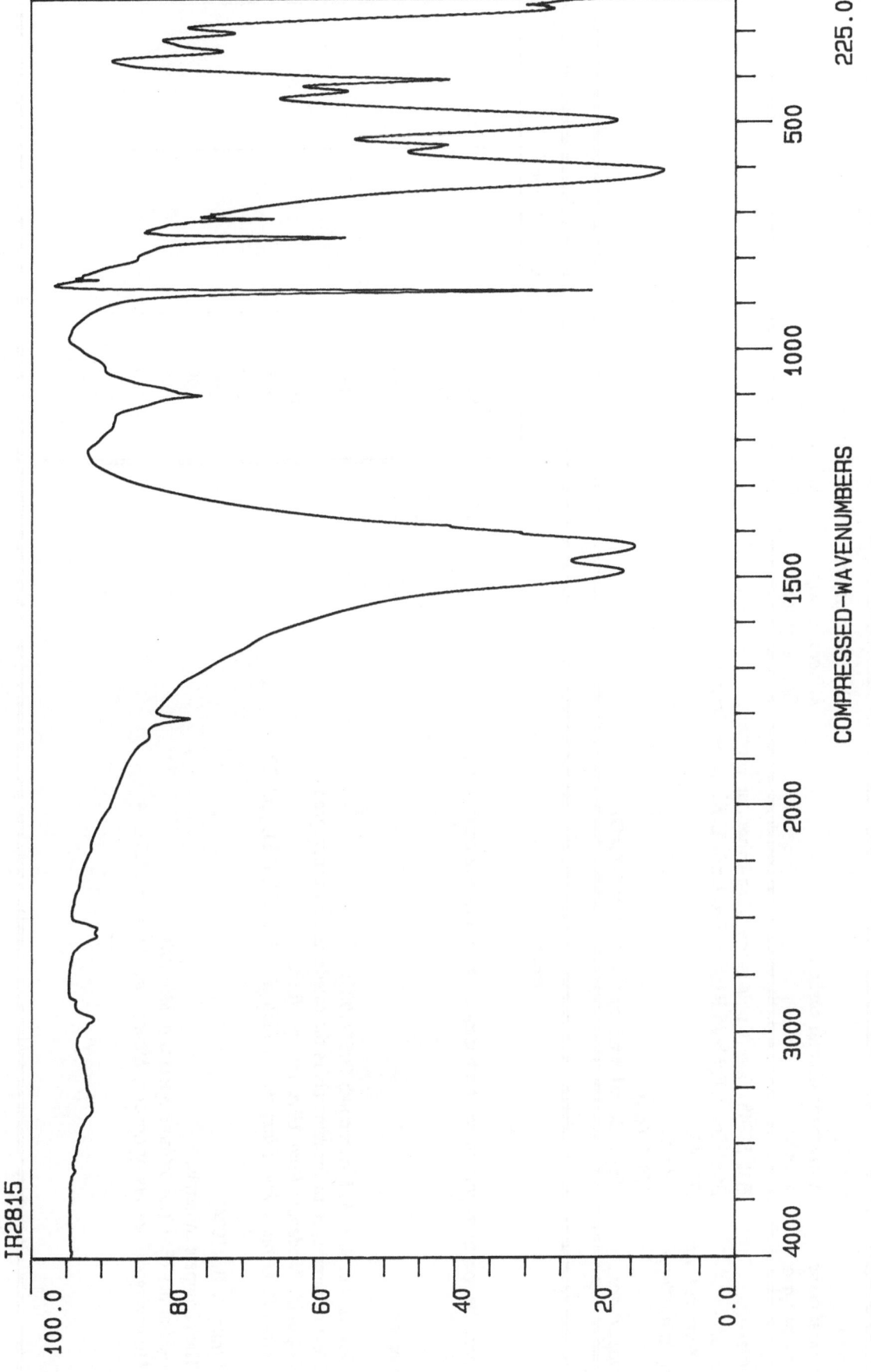

IR2815

% TRANSMITTANCE

COMPRESSED-WAVENUMBERS

STENONITE

STRONTIANITE

Formula:	SrCO₃	Crystal system:	Orthorhombic
Chemical class:	Anhydrous normal carbonate	Mineral group:	Aragonite
Chemical type:	A(XO₃)	Space group:	Pmcn

Specimen:	BM 58848 Large pale green fibro-columnar crystals.
Source:	Strontian, Highland Region, Scotland, U.K. (type locality).
Spectrum ref. no.:	IR 2672
Sample medium:	KBr disk
XRD:	1704F (std)
Composition:	Major Sr with minor Ca (3·8 wt% CaO)

Notes

Compare the spectrum with those of other members of the aragonite group.

References:

1. Gevork'yan S.V. & Povarennikh O.S. (1983)
   New infrared spectra for minerals in the calcite and aragonite groups.
   Dopovidi Akademiyi Nauk Ukrayins'koyi RSR,
   Seriya B: Geologichni, Khimichni ta Biologichni Nauki, (11), pp.8-12.

2. White W.B. (1974)
   The carbonate minerals.
   In: Farmer (Ed) *The Infrared Spectra of Minerals.*
   Mineralogical Society of London, Monograph No. 4, pp.227-284.

Peak Table cm⁻¹

[3313]
2925
2877
2603
2487
1774
1458
1385
1073
858
843
706
699
670
520
467
242?

IR2672

% TRANSMITTANCE

COMPRESSED-WAVENUMBERS

STRONTIANITE

100.0   80   60   40   20   0.0

4000   3000   2000   1500   1000   500   225.0

STRONTIODRESSERITE

Formula:	$(Sr,Ca)Al_2(CO_3)_2(OH)_4 \cdot H_2O$
Chemical class:	Hydrated carbonate with hydroxyl or halogen
Chemical type:	$A_mB_n(XO_3)_pZ_q \cdot xH_2O$ with $(m+n):p = 3:2$

Crystal system:	Orthorhombic
Mineral group:	Alumohydrocalcite (dundasite)
Space group:	Pbmm

Specimen:	BM 1983,643 Spheroidal aggregates of white acicular crystals.
Source:	Francon Quarry, St Michel, Montreal Island, Quebec, Canada. (Type locality)
Spectrum ref. no.:	IR2881
Sample medium:	KBr disk
XRD:	8205F (std)
Composition:	Sr:Ca:Al ≈ 1·0·1:2 with trace Ba & Na

Peak Table cm⁻¹

3591	1090
3508	1066
3471	965
3175	887
3075	849
2606	841
2457	759
2241	743
2151	726
2087	682
1859	577
1805	552
1645	481
1561	455
1509	383
1457	311
1373	278
1110	

Notes

The spectrum is similar to, but distinguishable from that of dresserite.
See ref.2 for peak assignments and comparison with dresserite and dundasite.

References:

1. Jambor J.L., Sabina A.P., Roberts A.C. & Sturman B.D. (1977)
Strontiodresserite, a new Sr Al carbonate from Montreal Island, Quebec.
Canadian Mineralogist, 15(3), pp.405-407.

2. Farrell D.M. (1977)
Infrared investigation of basic double carbonate hydrate minerals.
Canadian Mineralogist, 15(3), pp.408-413.

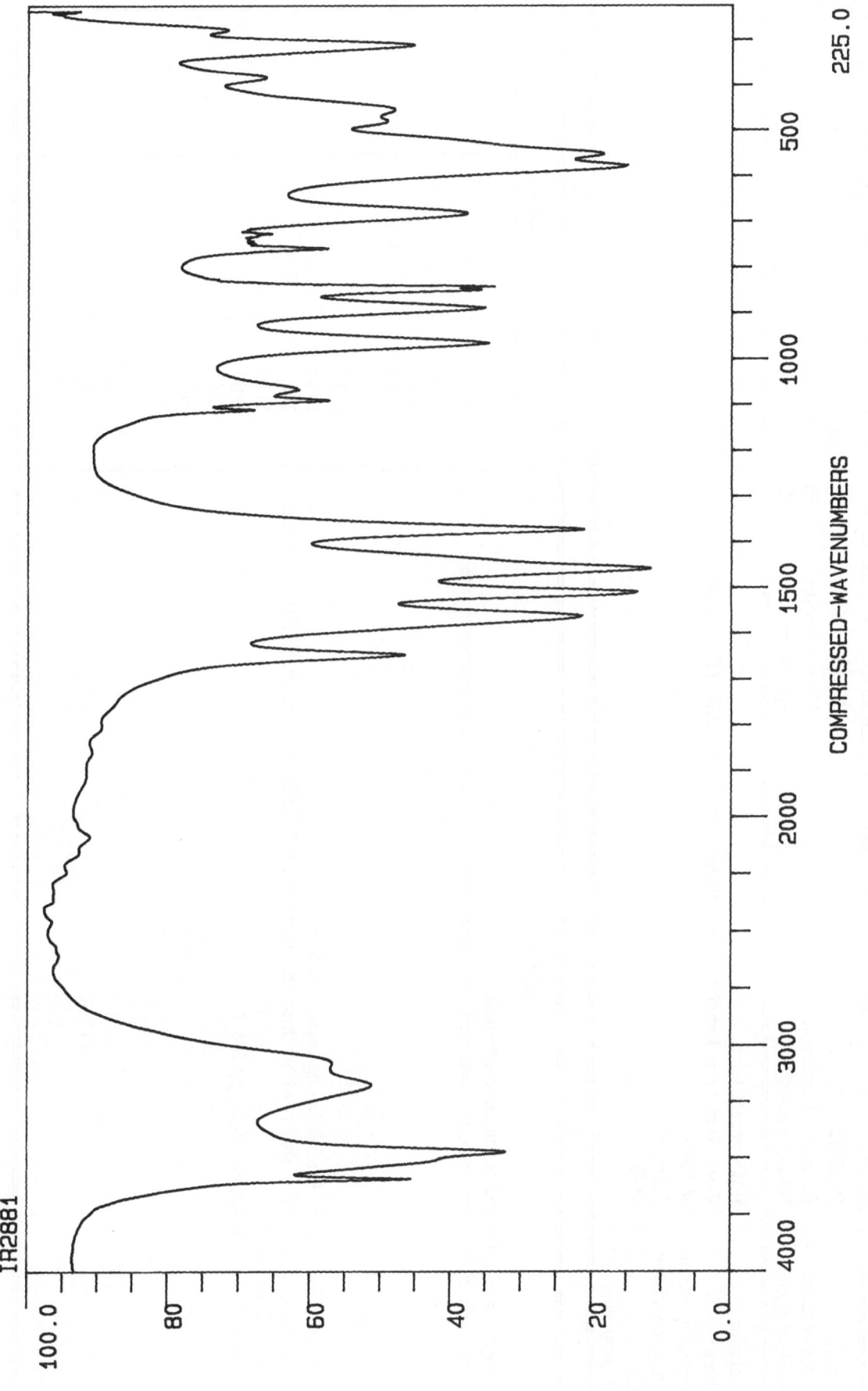

IR2881

% TRANSMITTANCE

COMPRESSED-WAVENUMBERS

STRONTIODRESSERITE

SUSANNITE

Crystal system:	**Trigonal**
Mineral group:	**Susannite**
Space group:	**R3̄**

Formula:	**Pb$_4$(SO$_4$)(CO$_3$)$_2$(OH)**
Chemical class:	**Compound carbonate**
Chemical type:	**Miscellaneous**
Specimen:	**RMS unregistered.**
Source:	**Roan Burn vein, Leadhills, Lanarkshire, Scotland, U.K.** (Type locality).
Spectrum ref. no.:	**IR2930**
Sample medium:	**KBr disk**
XRD:	**9998**
Composition:	

Peak Table cm^{-1}

3577
3431
2959
2924
2856
2411
1734
1628
1403
1116
1080
1051
964
840
706
682
602
421

Notes

Trimorphous with leadhillite and macphersonite.
The spectrum is very similar to, but distinguishable from, those of leadhillite and macphersonite.

References:

1. Russell J.D., Fraser A.R. & Livingstone A. (1984)
   The infrared absorption spectra of the three polymorphs of PbSO$_4$(CO$_3$)$_2$(OH)$_2$ (leadhillite, susannite and macphersonite).
   Mineralogical Magazine, **48**(2), pp.295-7.

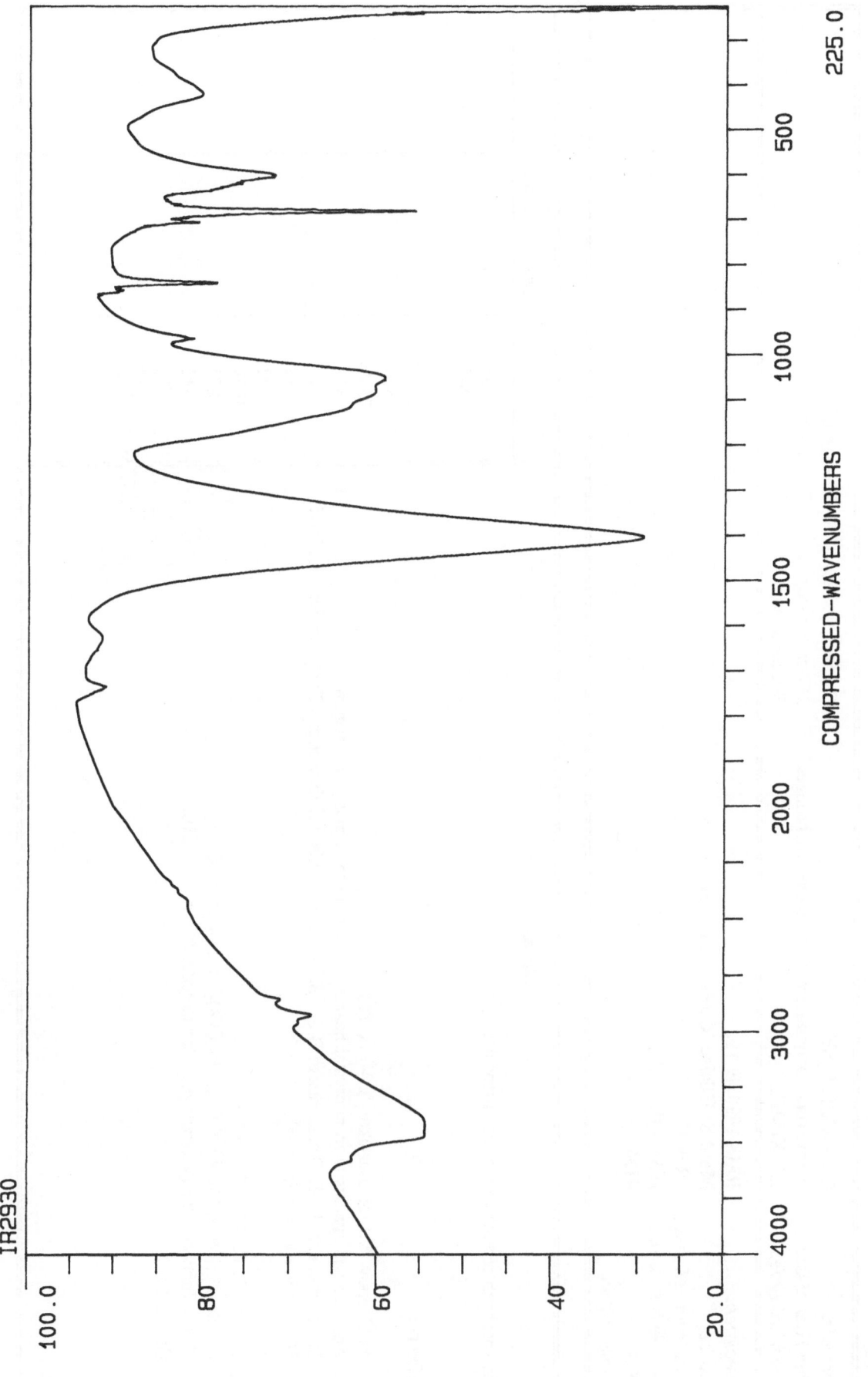

SUSANNITE

SYNCHYSITE-(Y) (doverite)

Formula:	$Ca(Y,Ce)(CO_3)_2F$
Chemical class:	Anhydrous carbonate with hydroxyl or halogen
Chemical type:	$AB(XO_3)Z_q$
Specimen:	RMS 1980.49.11.
Source:	Mont St Hilaire, Quebec, Canada.
Spectrum ref. no.:	IR2948
Sample medium:	KBr disk
XRD:	4156
Composition:	

Crystal system:	Hexagonal
Mineral group:	Bastnäsite
Space group:	?

Peak Table cm⁻¹

3687
3440
2925
2860
2500
2349
1816
1744
1464
1482
1079
871
741
602
352
287

Notes

The spectrum resembles that of bastnäsite.

References:

1. Yukhtanov P.P. & Burlakov Y.V. (1985)
   Ankylite and synchisite from crystal bearing pockets in the Polar Urals region
   In: Yushkin, N.P. & Ostashchenko B.A. *Minerals and mineral formation. Trudy Institut Geologii.*
   50, pp.99-104. (In Russian).

2. Scharm B. & Kühn P. (1983)
   Synchisite; (Nd), Ca(Nd,Y,Gd,) [F/(CO$_3$)$_2$], a new mineral.
   Neues Jahrbuch für Mineralogie, Monatshefte, (5), pp.201-210.
   (In English).

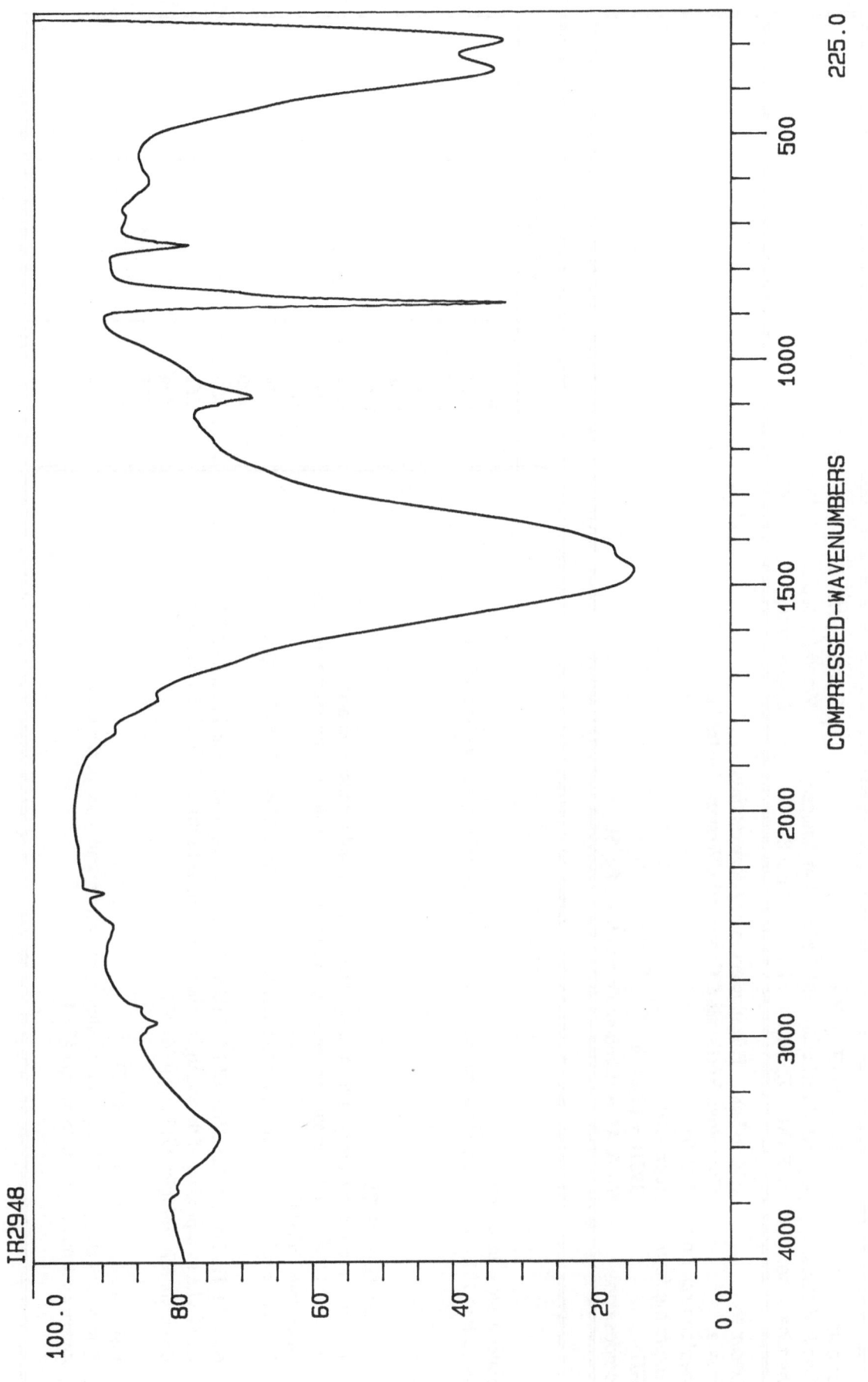

IR2948

% TRANSMITTANCE

100.0 80 60 40 20 0.0

4000 3000 2000 1500 1000 500 225.0

COMPRESSED-WAVENUMBERS

SYNCHISITE- (Y)

TAKOVITE

Formula:	$Ni_6Al_2(CO_3)(OH)_{16}\cdot4H_2O$
Chemical class:	Hydrated carbonate with hydroxyl or halogen
Chemical type:	$A_mB_n(XO_3)_pZ_q\cdot xH_2O$ with $(m+n):p = 8:1$
Crystal system:	Trigonal
Mineral group:	Pyroaurite
Space group:	$R\bar{3}m$

Specimen:	BM 1976,81 Pale blue/green powdery coating.
Source:	Carr Boyd Rocks mine, Goongarrie, Western Australia.
Spectrum ref. no.:	IR2804
Sample medium:	KBr disk
XRD:	19310 = takovite
Composition:	Ni & Al with minor Zn and trace Fe, Si

Peak Table cm^{-1}

3418
1734
1617
1560
1396
1351
1282
1015
910
819
690
623
561
434
365
333
254

Notes

Former name = eardleyite

The peak at 1015 cm^{-1} is thought to be due to inseperable kaolinite impurity - see ref. no.3.

References:

1. Brindley G.W. (1978)
   The structure and chemistry of the hydrous nickel silicate and aluminate minerals.
   In: Goni J.(Ed), Colloque sur la mineralogie, geochimie, geologie des mineraux et minerais nickeliferes lateritiques,
   Fr., Bur. Rech. Geol. Minieres, Bull.,(Ser.2),Sect 2, Geol.Gites Miner., **3**, pp.233-45.

2. Nickel E.H., Davis C.E.S., Bussell M., Bridge P.J., Dunn J.G. & MacDonald R.D. (1977)
   Eardleyite as a product of the supergene alteration of nickel sulfides in Western Australia.
   American Mineralogist. **62**(5,6), pp.449-57.

3. Bish D.L. & Brindley G.W. (1977)
   A reinvestigation of takovite, a nickel aluminum hydroxy carbonate of the pyroaurite group.
   American Mineralogist, **62**(5,6), pp.458-64.

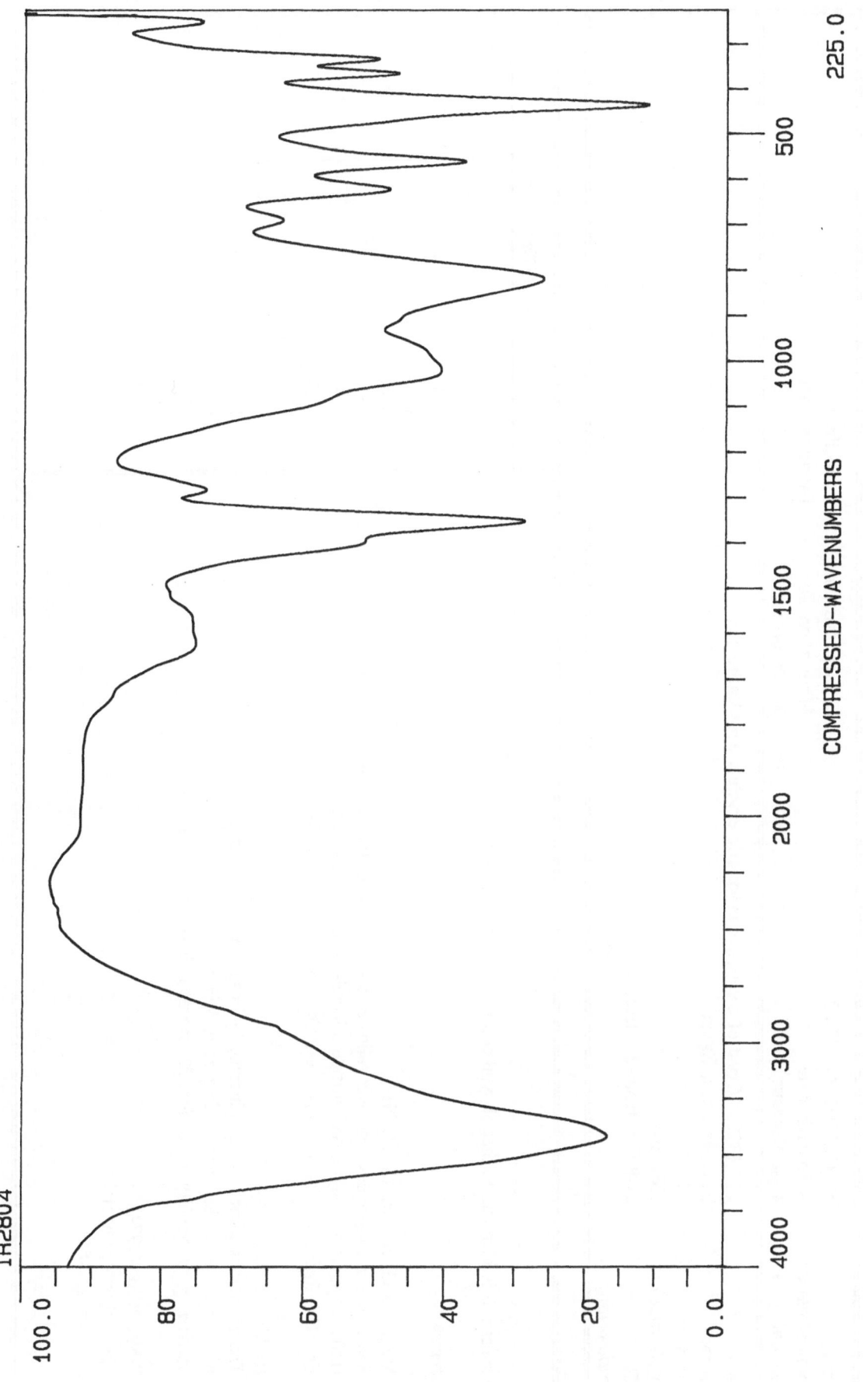

IR2804

% TRANSMITTANCE

COMPRESSED-WAVENUMBERS

TAKOVITE

TRONA

Formula:	Na$_3$(CO$_3$)(HCO$_3$)·2H$_2$O
Chemical class:	Acid carbonate
Chemical type:	Miscellaneous

Crystal system:	Monoclinic
Mineral group:	Thermonatrite
Space group:	I2/a

Specimen:	BM 59235 Crust of colourless prismatic crystals with halite etc.
Source:	Natron Lakes, Egypt.
Spectrum ref. no.:	IR2876
Sample medium:	KBr disk
XRD:	8149F = trona (+ trace quartz).
Composition:	

Peak Table cm^{-1}

3469
3067
2536
2444
2266
1692
1465
1357
1191
1113
1065
1015
851
651
601

Notes

The spectrum is identical to that of synthetic sodium sesquicarbonate.

References:

1. Maglione G. & Carn M. (1975)
    Spectres infrarouges des mineraux salins et des silicates neoformes dans le Bassin tchadien.
    (Infrared spectra of saline and silicate minerals from the Chad Basin).
    Fr., Off. Rech. Sci. Tech. Outre Mer, Cah., Ser. Geol. 7, pp.3-9

2. Ryskin Ya. I. (1974)
    The vibrations of protons in minerals; hydroxyl, water and ammonium.
    In: Farmer (Ed) *The Infrared Spectra of Minerals*,
    Mineralogical Society of London, Monograph No.4, pp.137-181.

3. White W.B. (1974)
    The carbonate minerals.
    Ibid. pp.227-84.

IR2876

% TRANSMITTANCE

COMPRESSED-WAVENUMBERS

TRONA

TUNISITE

Formula:	NaCa$_2$Al$_4$(CO$_3$)$_4$(OH)$_8$Cl
Chemical class:	Anhydrous carbonate with hydroxyl or halogen
Chemical type:	Miscellaneous

Crystal system:	Tetragonal
Mineral group:	Dawsonite
Space group:	P4/nmm

Specimen:	BM 1981,482 Creamy white bladed crystals.
Source:	Condorcet, Drôme, France.
Spectrum ref. no.:	IR2791
Sample medium:	KBr disk
XRD:	7959F = tunisite
Composition:	Na:Ca:Al:Cl ≈ 0·8:2·1:4:1

Peak Table cm^{-1}

3495	674
3459	533
3410	468
2926	416
2603	384
2279	311
1998	274
1915	
1866	
1563	
1513	
1475	
1156	
1131	
1091	
981	
844	
797	
741	

Notes

The spectrum matches that given in Suhner, (5-57) tunisite.

References:

1. Effenberger H., Kluger F., Pertlik F. & Zemann J. (1981)
    Tunisite; crystal structure and revision of chemical formula.
    Tschermaks Mineralogische und Petrographische Mitteilungen, **28**(1), pp.65-77.
    (In German with English summary)

2. Zdenek J., Povondra P. & Ervin S. (1969)
    Tunisite, a new carbonate from Tunisia.
    American Mineralogist, **54**(1,2), pp.1-13.

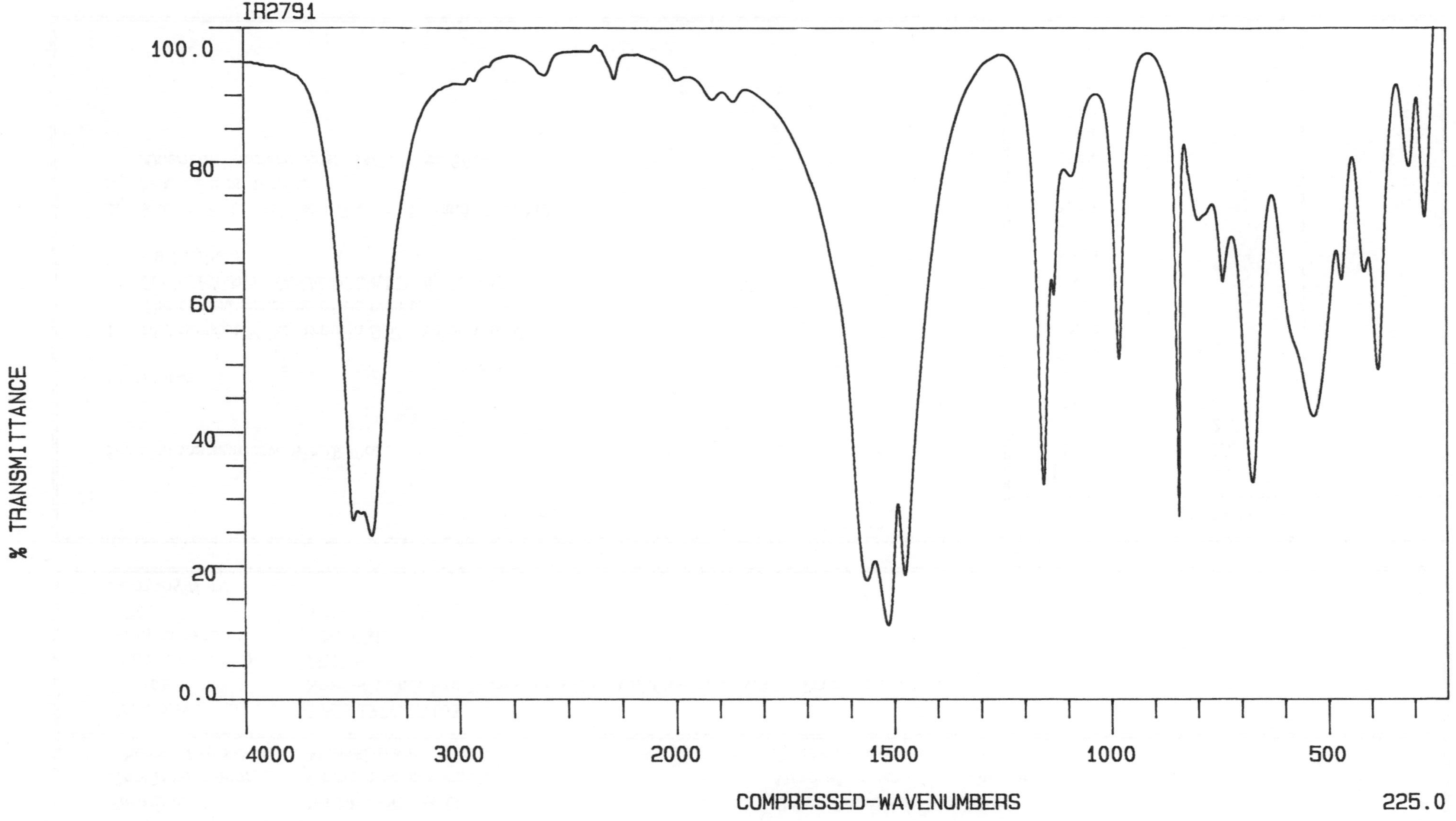

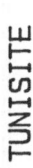

IR2791

% TRANSMITTANCE

100.0   80   60   40   20   0.0

4000   3000   2000   1500   1000   500   225.0

COMPRESSED-WAVENUMBERS

TUNISITE

TYCHITE

Formula:	Na$_6$Mg$_2$ (SO$_4$)(CO$_3$)$_4$
Chemical class:	Compound carbonate
Chemical type:	Miscellaneous
Crystal system:	Cubic
Mineral group:	Susannite
Space group:	Fd3

Specimen:	RMS 1974.47.165
Source:	Searles Lake, San Bernardino Co., California, U.S.A. (Type locality)
Spectrum ref. no.:	IR2924
Sample medium:	KBr disk
XRD:	4103
Composition:	

Peak Table cm^{-1}

[3417]	396
2924	333
2859	273
2649	
2539	
2339	
1818	
1731	
1630	
1463	
1447	
1385	
1110	
1104?	
884	
859	
718	
662	
631	

Notes

See also **manganotychite** IR3062.

References:

1. Malinovskii Y.A., Baturin S.V. & Belov N.V. (1979)
   The crystal structure of Fe tychite.
   Soviet Physics. Doklady. **24**(12), pp.951-3.
   (In English)

2. Keester K.L., Johnson G.G.Jr. & Vand V. (1969)
   New data on tychite.
   American Mineralogist, **54**(1,2), pp.302-5.

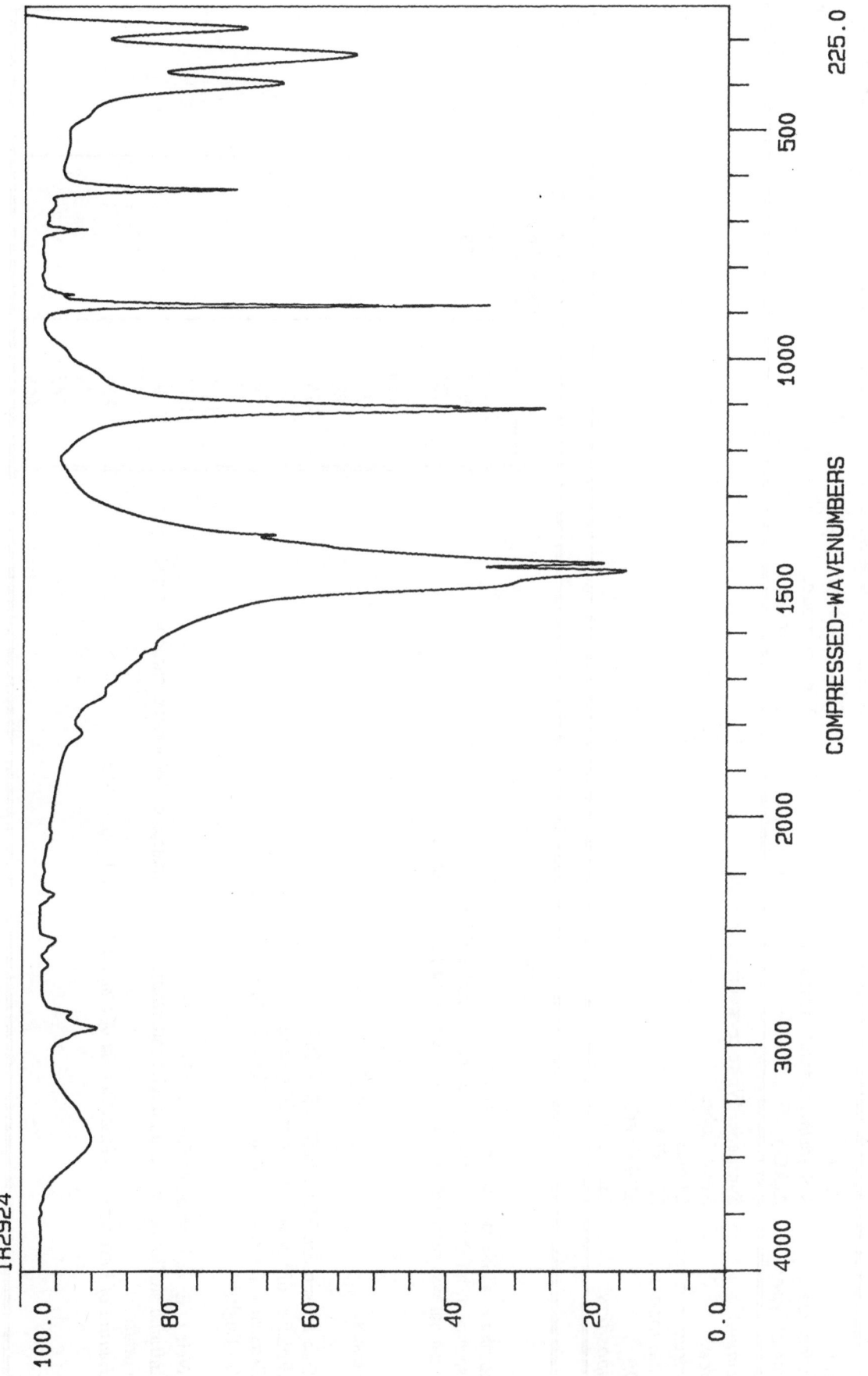

IR2924

% TRANSMITTANCE

COMPRESSED-WAVENUMBERS

TYCHITE

VATERITE

Formula:	CaCO$_3$
Chemical class:	Anhydrous normal carbonate
Chemical type:	A(XO$_3$)

Crystal system:	Hexagonal
Mineral group:	Calcite
Space group:	P6$_3$/mmc

Specimen:	Synthetic. White crystalline powder.
Source:	NHM Labs.
Spectrum ref. no.:	IR2603
Sample medium:	KBr disk
XRD:	6202F (std)
Composition:	Ca only

Peak Table cm^{-1}

[3228]	341
2973	291
2905	
2624	
2507	
2356	
1836	
1765	
1743	
1489	
1432	
1408	
1089	
999	
877	
850	
844	
745	
668	

Notes

Trimorphous with **aragonite** and **calcite**.
The spectra of all three polymorphs are easily distinguished.
The material was prepared according to the method given in ref.1.

References:

1. Sato Mitsuo and Matsuda Shunji. (1969)
Structure of vaterite and infrared spectra.
Zeitschrift für Kristallographie, **129**(5,6), pp.405-10.
(In English)

2. Weir C.E. & Lippincott E.R. (1961)
Infrared studies of aragonite, calcite and vaterite type structures in the borates, carbonates and nitrates.
Journal of Research, National Bureau of Standards, **65**(A), pp.173-83.

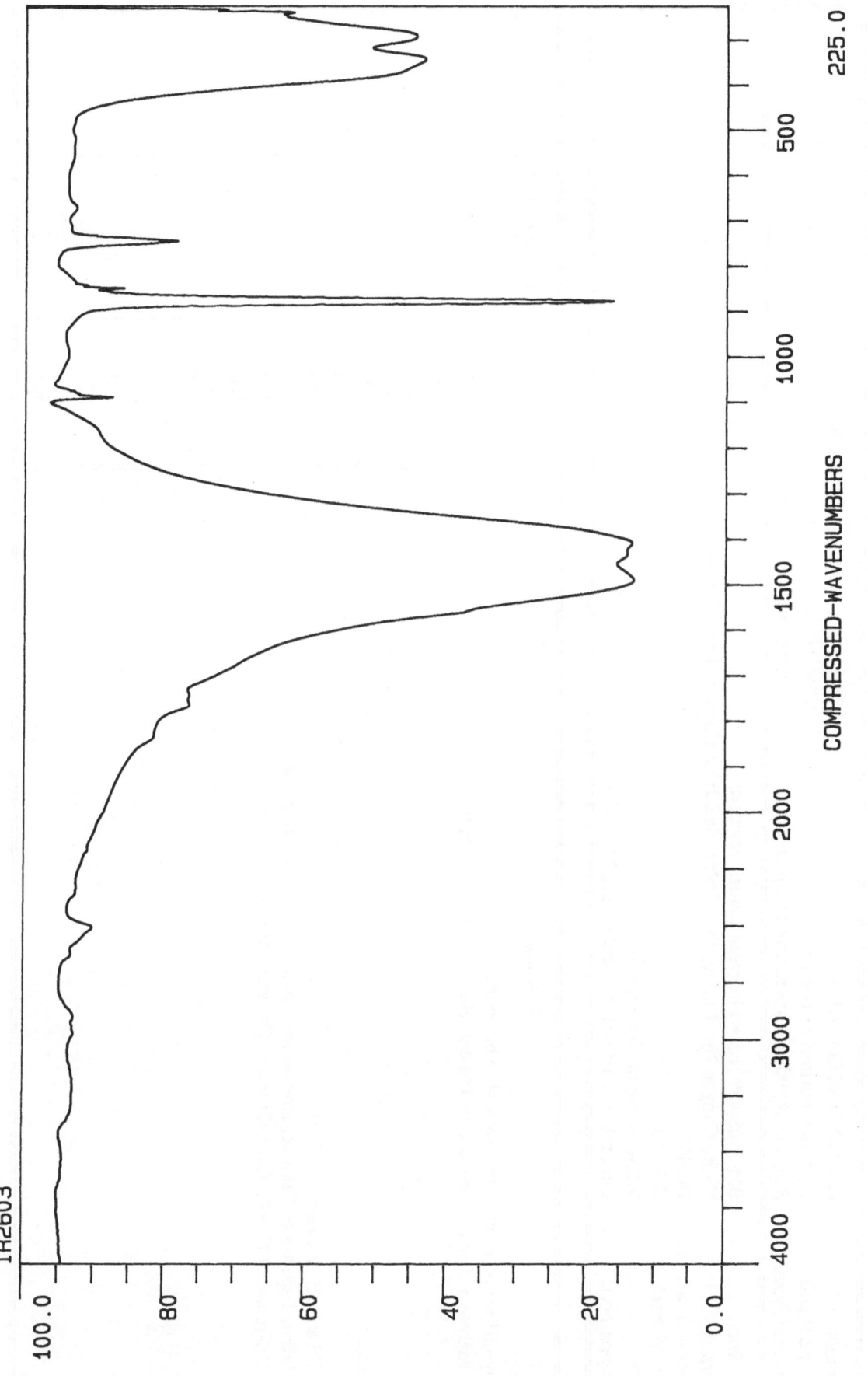

IR2603

% TRANSMITTANCE

COMPRESSED-WAVENUMBERS

VATERITE   synthetic

VOGLITE

Formula:	$Ca_2Cu(UO_2)(CO_3)_4 \cdot 6H_2O$
Chemical class:	Hydrated normal carbonate
Chemical type:	$A_m B_n(XO_3)_p \cdot xH_2O$ where (m+n) : p > 1 : 1
Specimen:	BM 1965,439 Emerald green bladed crystals.
Source:	White Canyon No. 1 mine, Frey Point, San Juan County, Utah, U.S.A.
Spectrum ref. no.:	IR2867
Sample medium:	KBr disk
XRD:	8123F = voglite (see notes)
Composition:	Ca:Cu:U ≈ 1·1:0·2:1·0 ? with trace Mg, S

Crystal system:	Monoclinic
Mineral group:	
Space group:	$P2_1$

Peak Table cm⁻¹

3411
2595
1563
1515
1429
1149
1114
1084
1025
904
838
794
746
717
671
600
523
474
298

Notes

X-ray diffraction showed the material to be poorly crystalline.

The analysis indicates a low copper content relative to the ideal formula.

References:

1. Urbanec Z & Čejka J. (1979)
   Infrared spectra of liebigite, andersonite, voglite, and schroeckingerite.
   Collection of Czechoslovak Chemical Communications, **44**(1), pp.10-23.

IR2867

% TRANSMITTANCE

100.0 80 60 40 20 0.0

4000 3000 2000 1500 1000 500 225.0

COMPRESSED-WAVENUMBERS

VOGLITE

WELOGANITE

Formula:	$Sr_3Na_2Zr(CO_3)_6 \cdot 3H_2O$
Chemical class:	**Hydrated normal carbonate**
Chemical type:	$A_mB_n(XO_3)_p \cdot xH_2O$ where $(m+n){:}p > 1{:}1$

Crystal system:	**Triclinic, pseudotrigonal**
Mineral group:	**McKelveyite**
Space group:	**P1**

Specimen:	**RMS 1976.34.1 Straw yellow hexagonal crystals.**
Source:	**Francon Quarry, St Michel, Montreal Island, Quebec, Canada. (Type locality).**
Spectrum ref. no.:	**IR2946**
Sample medium:	**KBr disk**
XRD:	**39476**
Composition:	

Peak Table cm⁻¹

3383	**672?**
3317	**547**
2931	**325**
2617	
2421	
1682	
1611	
1555	
1527	
1413	
1354	
1064	
1057	
869	
850	
761	
749	
706	
678	

Notes

Forms a series with **donnayite-(Y)**.

The spectrum closely matches that shown in the original description, ref. 1, and is similar to those of **donnayite-(Y)** and **mckelveyite-(Y)**.

References:

1. Sabina A.P., Jambor J.L. & Plant A.G. (1968)
   Weloganite, a new strontium zirconium carbonate from Montreal Island, Canada.
   Canadian Mineralogist, 9(4), pp. 468-77.
   with correction, *Canadian Mineralogist*, 1969, 9(5), p654.

IR2946

% TRANSMITTANCE

COMPRESSED-WAVENUMBERS

WELOGANITE

WITHERITE

Formula:	**BaCO$_3$**
Chemical class:	**Anhydrous normal carbonate**
Chemical type:	**A(XO$_3$)**
Specimen:	**BM 26683** Colourless prismatic crystal groups.
Source:	Fallowfield mine, Hexham, Northumbria, County Durham, U.K.
Spectrum ref. no.:	**IR2671**
Sample medium:	**KBr disk**
XRD:	
Composition:	**Ba only**

Crystal system:	**Orthorhombic**
Mineral group:	**Aragonite**
Space group:	**Pmcn**

Peak Table cm^{-1}

[3444]
2878
2821
2541
2452
2093
1752
1431
1060
858
840
709
693
668
390
308
286
242?

Notes

The spectrum is identical with those of samples from other localities.
Compare the spectrum with those of other members of the **aragonite** group.

References:

1. Anatassov V., Vassileva M. & Goranova R. (1989)
   Carbonates with aragonite type structure (aragonite, witherite and cerussite) in the Kremikotvsi deposit. (Hungarian with English abstract).
   Annual of the Higher Institute of Mining & Geology, Sofia, Part 1, Geology, **35**.

2. White W B. (1974)
   The carbonate minerals.
   In: Farmer (Ed) *The Infrared Spectra of Minerals*
   Mineralogical Society of London, Monograph No.4, pp.227-84.

3. Decius J.C., Malan O.G. & Thompson H.W. (1963)
   The effects of intermolecular forces upon the vibration of molecules in the crystalline state. 1. The out-of-plane bending of the carbonate ion in aragonite minerals.
   Proceedings of the Royal Society of London, Series A, **275**, pp.295-309.

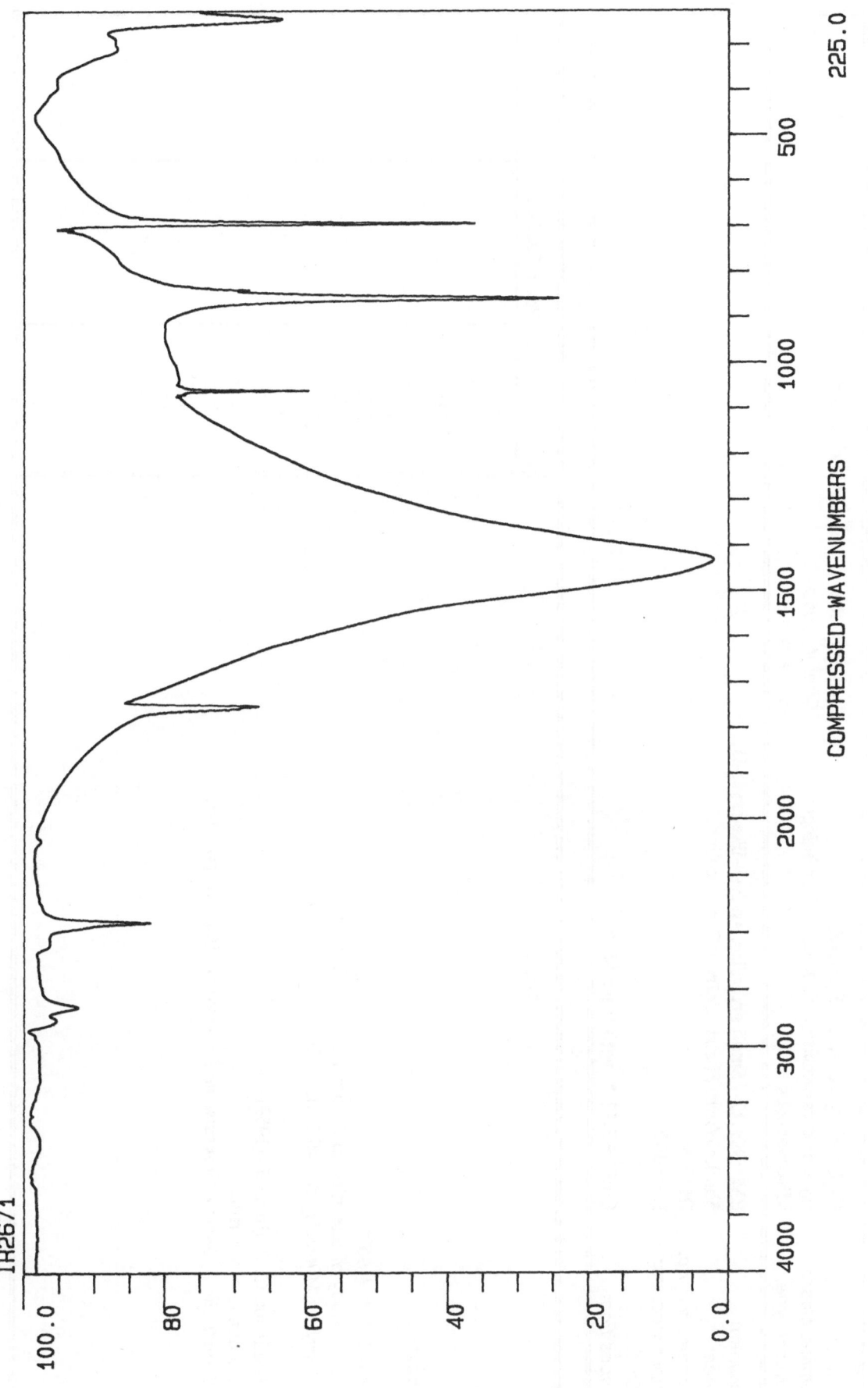

IR2671

% TRANSMITTANCE

COMPRESSED—WAVENUMBERS

WITHERITE

WYARTITE

Formula:	$Ca_3U^{4+}(UO_2)_6(CO_3)_2(OH)_{18}\cdot 4H_2O$
Chemical class:	Hydrated carbonate with hydroxyl or halogen
Chemical type:	Miscellaneous

Crystal system:	Orthorhombic
Mineral group:	
Space group:	$P2_12_12_1$

Specimen:	BM 1969,47 Small green/brown lath-like crystals.
Source:	Shinkolobwe, Shaba, Zaïre. (Type locality).
Spectrum ref. no.:	IR2866
Sample medium:	KBr disk
XRD:	
Composition:	Ca:U ≈ 3:11 ? with trace Si, S

Notes

References:

1. Clark J.R. (1960)
   X-ray study of alteration in the uranium mineral wyartite.
   American Mineralogist, **45**(1,2), pp.200-8.

2. Guillemin C. & Protas J. (1959)
   Lanthinite et wyartite.
   Bulletin de la Société Francaise de Minéralogie, **82**(1,3), pp.80-6.

Peak Table cm⁻¹

3417
2937
2861
1745
1623
1535
1381
1372
1165
1054
902
795
742
571
453
371
274

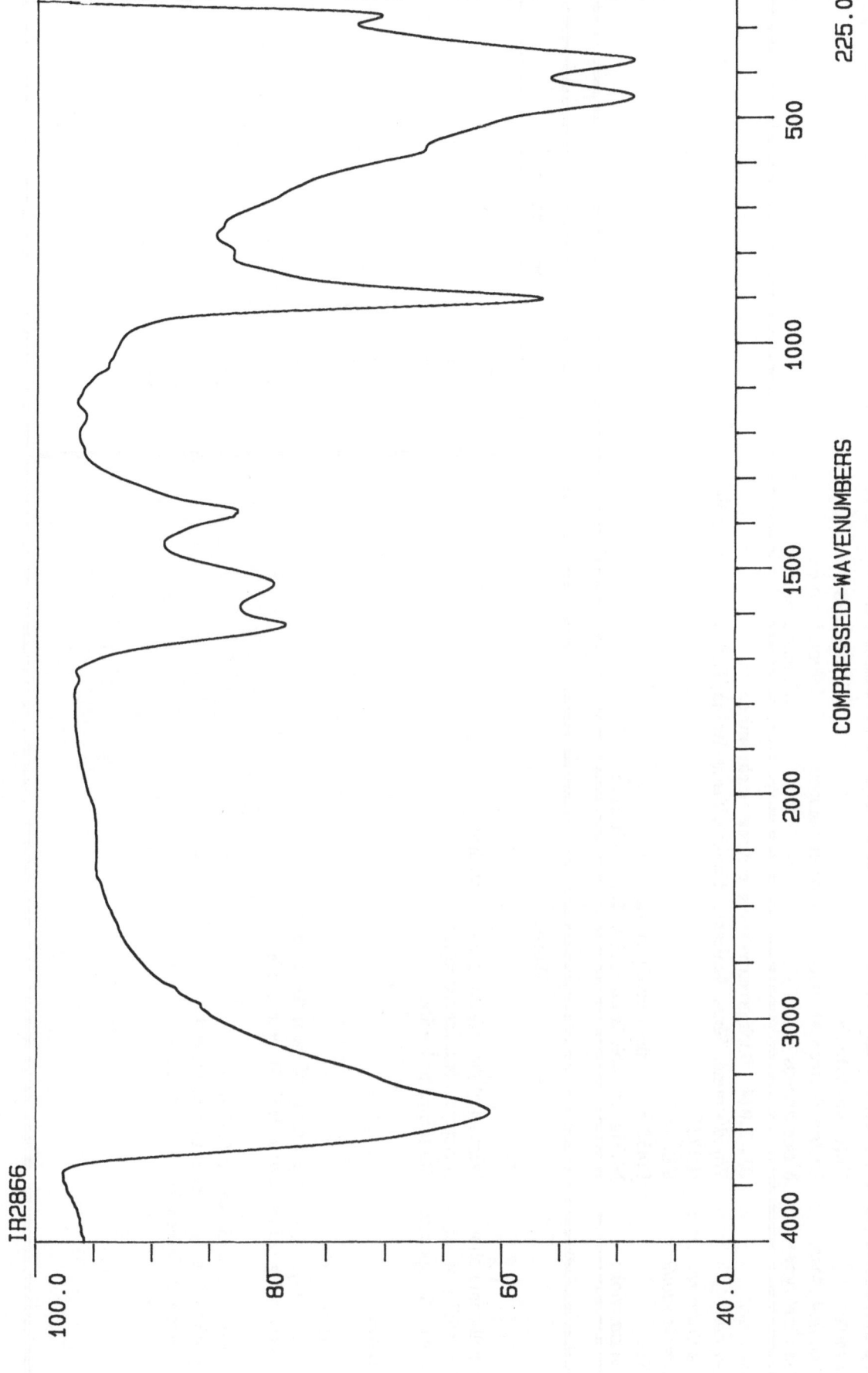

IR2866

% TRANSMITTANCE

COMPRESSED-WAVENUMBERS

WYARTITE

ZARATITE

Formula:	$Ni_3(CO_3)(OH)_4 \cdot 4H_2O$
Chemical class:	Hydrated carbonate with hydroxyl or halogen
Chemical type:	Miscellaneous
Crystal system:	Cubic
Mineral group:	
Space group:	?

Specimen:	BM 22014 Dark green vitreous coating on chromite.
Source:	Wood's mine, Texas, Lancaster County, Pennsylvania, U.S.A. (Type locality).
Spectrum ref. no.:	IR2735
Sample medium:	KBr disk
XRD:	12438 = zaratite (poorly crystalline).
Composition:	Ni:Mg variable from 11:1 to 16:1 with trace Si, S

Peak Table cm⁻¹

3526
3426
2927
2860
1580
1384
1070
1022
985
873
834
678
522
438
400

Notes

Partially amorphous, as were all of the zaratite specimens studied.
The validity of zaratite as a species is discussed in ref. 1.
Compare the spectrum with that of **hellyerite**.

References:

1. Isaacs T. (1963)
    The mineralogy and chemistry of the nickel carbonates.
    Mineralogical Magazine, **33**(263), pp.663-678.

2. Huang C.K. & Kerr P.F. (1960)
    Infrared study of the carbonate minerals.
    American Mineralogist, **45**, pp.311-24.

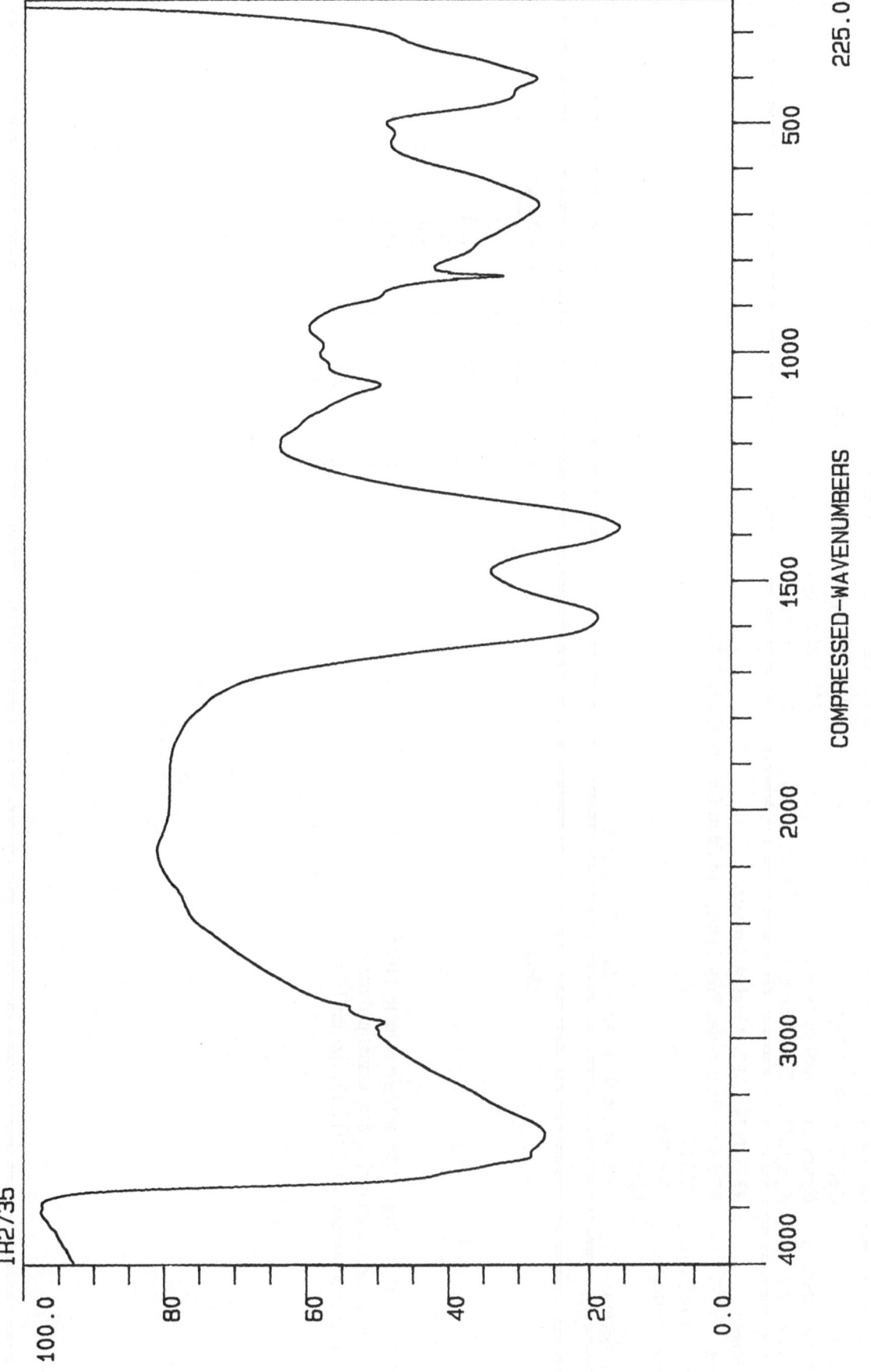

ZARATITE

ZELLERITE

Formula:	$Ca(UO_2)(CO_3)_2 \cdot 5H_2O$
Chemical class:	Hydrated normal carbonate
Chemical type:	$A_m B_n (XO_3)_p \cdot xH_2O$ where $(m+n):p = 1:1$

Crystal system:	Orthorhombic
Mineral group:	
Space group:	$Pmn2_1$

Specimen:	RMS 1978.17.98. Yellow fibrous.
Source:	White Canyon mine, Frey Point, San Juan Co., Utah, U.S.A.
Spectrum ref. no.:	IR2917
Sample medium:	KBr disk
XRD:	3897
Composition:	Ca:U = 1:0·9 + trace Mg,Sr,Fe,Mn,Ba

Peak Table cm^{-1}

3556	772?
3418	754
2925	741?
2856	700
1778	692?
1636	621
1520	309
1439	249?
1429	
1378	
1167	
1090	
963	
953?	
924	
857?	
843?	
830	
822?	

Notes

References:

1. Coleman R.G., Ross D.R. & Meyrowitz R. (1966)
   Zellerite and metazellerite, new uranyl carbonates.
   American Mineralogist, **51**(11,12), pp.1567-78.

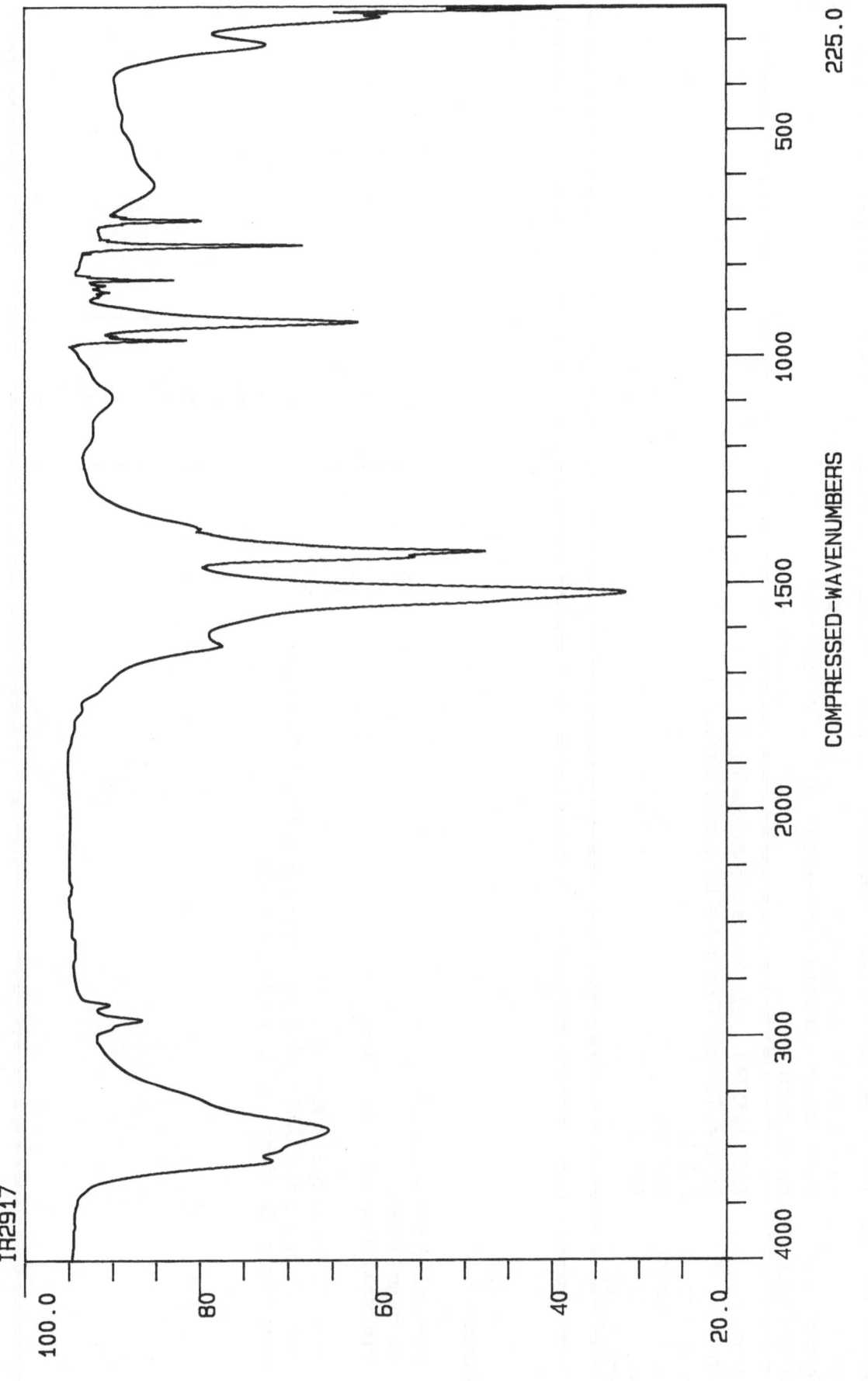

IR2917

% TRANSMITTANCE

100.0   80   60   40   20.0

4000   3000   2000   1500   1000   500   225.0

COMPRESSED-WAVENUMBERS

ZELLERITE

ZNUCALITE

Formula:	$Zn_{12}Ca(UO_2)(CO_3)_3(OH)_{22} \cdot 4H_2O$	Crystal system:	Triclinic
Chemical class:	Hydrated carbonate with hydroxyl or halogen	Mineral group:	
Chemical type:	Miscellaneous	Space group:	P1

Specimen:	RMS 1992.49.1 yellow/cream coloured coating.
Source:	Příbram, Bohemia, Czechoslovakia. (type locality).
Spectrum ref. no.:	IR2923
Sample medium:	KBr disk
XRD:	4547
Composition:	

Peak Table cm⁻¹

3331
2965
2930
1734
1508
1392
1082
1046
950
890
832
801
742
706
612
517
472
369
276

Notes

References:

1. Jambor J.L. & Puziewicz J. (1991)
   New mineral names.
   American Mineralogist, **76**, pp.1728-35.

2. Ondruš P., Veselovský F. & Rybka R. (1990)
   Znucalite, $Zn_{12}(UO_2)Ca(CO_3)(OH)_{22} \cdot H_2O$, a new mineral from Příbram, Czechoslovakia.
   Neues Jahrbuch für Mineralogie, Monatshefte, pp.393-400.

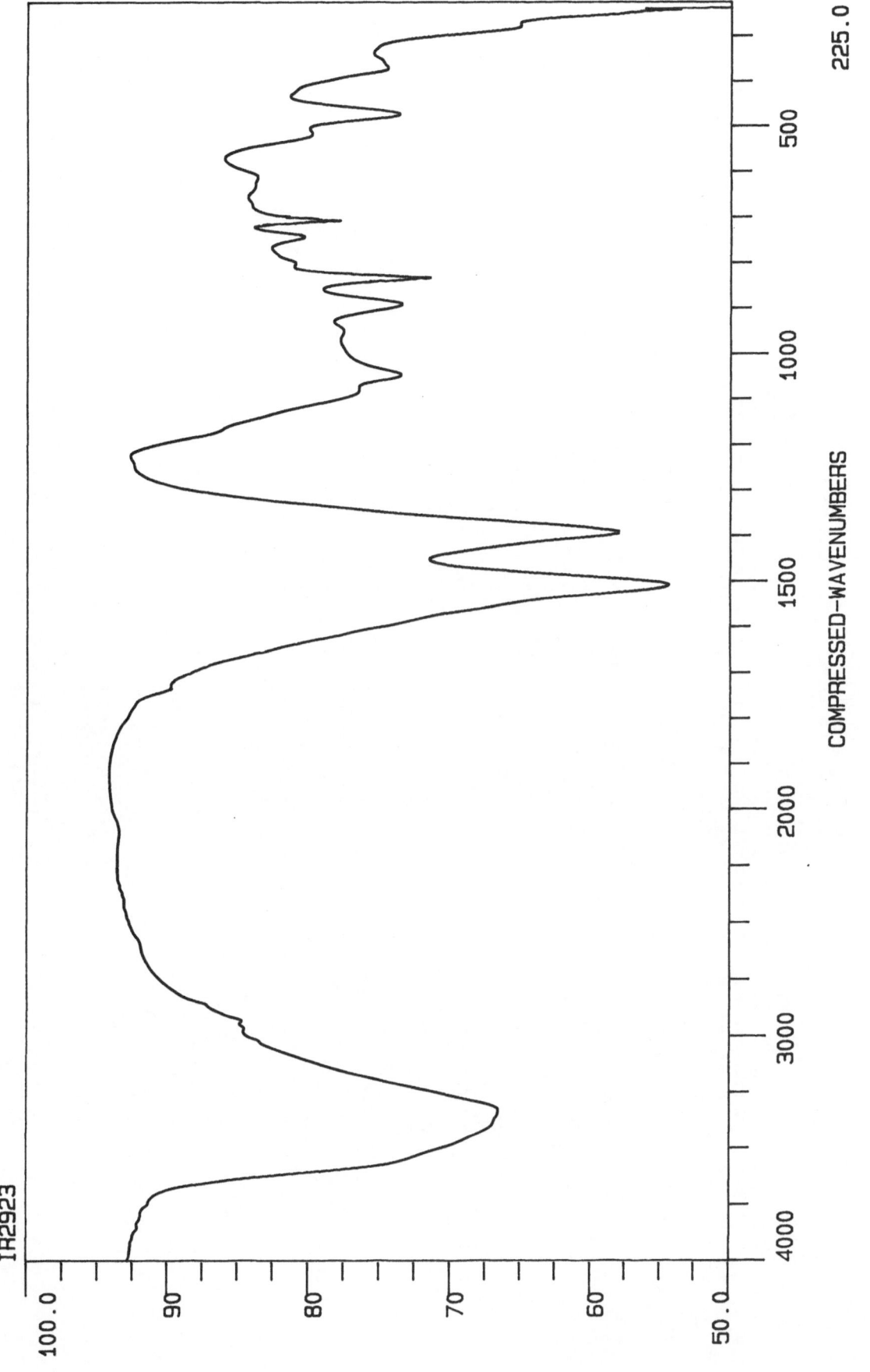

IR2923

% TRANSMITTANCE

COMPRESSED-WAVENUMBERS

ZNUCALITE